THE

# LOGIC AND UTILITY

OF

# MATHEMATICS,

WITH THE BEST METHODS OF INSTRUCTION EXPLAINED
AND ILLUSTRATED.

BY CHARLES DAVIES, LL.D.

NEW YORK:
PUBLISHED BY A. S. BARNES & CO.,
NO. 51 JOHN-STREET.
CINCINNATI:—H. W. DERBY & COMPANY.
1850.

Stereotyped by
RICHARD C. VALENTINE,
New York.

F. C. GUTIERREZ, Printer,
No. 51 John-street, corner of Dutch.

# PREFACE.

THE following work is not a series of speculations. It is but an analysis of that system of mathematical instruction which has been steadily pursued at the Military Academy over a quarter of a century, and which has given to that institution its celebrity as a school of mathematical science.

It is of the essence of that system that a principle be taught before it is applied to practice; that general principles and general laws be taught, for their contemplation is far more improving to the mind than the examination of isolated propositions; and that when such principles and such laws are fully comprehended, their applications be then taught as consequences or practical results.

This view of education led, at an early day, to the union of the French and English systems of mathematics. By this union the exact and beautiful methods of generalization, which distinguish the French school, were blended with the practical methods of the English system.

The fruits of this new system of instruction have been abundant. The graduates of the Military Academy have been sought for wherever science of the highest grade has been needed. Russia has sought them to construct her railroads;* the Coast Survey needed their aid; the works of internal improvement of the first class in our country, have mostly been conducted under their direction; and the recent war with Mexico afforded ample opportunity for showing the thousand ways in which science—the highest class of knowledge—may be made available in practice.

All these results are due to the system of instruction. In that system Mathematics is the basis—Science precedes Art—Theory goes before Practice—the general formula embraces all the particulars.

It was deemed necessary to the full development of the plan of the work, to give a general view of the subject of Logic. The materials of Book I. have been drawn, mainly, from the works of Archbishop Whately and Mr. Mill. Although the general outline of the subject has but little resemblance to the work of either author, yet very much has been taken from both; and in all cases where it could be done consistently with my own plan, I have adopted their exact language. This remark is particularly applicable to Chapter III., Book I., which is taken, with few alterations, from Whately.

For a full account of the objects and plan of the work, the reader is referred to the Introduction.

FISHKILL LANDING,
*June*, 1850.

* Major Whistler, the engineer, to whom was intrusted the great enterprise of constructing a railroad from St. Petersburg to Moscow, and Major Brown, who succeeded him at his death, were both graduates of the Military Academy.

# CONTENTS.

## INTRODUCTION.



## BOOK I.

## LOGIC.

### CHAPTER I.

## CHAPTER II.



## CHAPTER III.

# BOOK II.

## MATHEMATICAL SCIENCE.

### CHAPTER I.



### CHAPTER II.



### SECTION I.

## CHAPTER III.



## CHAPTER IV.

# BOOK III.

## UTILITY OF MATHEMATICS.

### CHAPTER I.



### CHAPTER II.



### CHAPTER III.



---

## APPENDIX.

# INTRODUCTION

## OBJECTS AND PLAN OF THE WORK.

Utility and Progress: Their influence in government: In education.

Utility and Progress are the two leading ideas of the present age. They were manifested in the formation of our political and social institutions, and have been further developed in the extension of those institutions, with their subduing and civilizing influences, over the fairest portions of a great continent. They are now becoming the controlling elements in our systems of public instruction.

What the basis of Utility and Progress.

What, then, must be the basis of that system of education which shall embrace within its horizon a Utility as comprehensive and a Progress as permanent as the ordinations of Providence, exhibited in the laws of nature, as made known by science? It must obviously be laid in the examination and analysis of those laws; and

Preparatory studies.

primarily, in those preparatory studies which fit and qualify the mind for such Divine Contemplations.

Bacon's Philosophy.

When Bacon had analyzed the philosophy of the ancients, he found it speculative. The great highways of life had been deserted. Nature, spread out to the intelligence of man, in all the minuteness and generality of its laws—in all the harmony and beauty which those laws develop—had scarcely been consulted by the ancient philosophers. They had looked within, and not without. They sought to rear systems on the uncertain foundations of human hypothesis and speculation, instead of resting them on the immutable laws of Providence, as manifested in the material world. Bacon broke the bars of this mental prison-house: bade the mind go free, and investigate nature.

Philosophy of the Ancients.

Foundations of Bacon's Philosophy:

Bacon laid the foundations of his philosophy in organic laws, and explained the several processes of experience, observation, experiment, and induction, by which these laws are made known. He rejected the reasonings of Aristotle because they were not progressive and useful; because they added little to knowledge, and contributed nothing to ameliorate the sufferings and elevate the condition of humanity.

Why opposed to Aristotle's.

Practical Philosophy:

The time seems now to be at hand when the philosophy of Bacon is to find its full development. The only fear is, that in passing from a speculative to a practical philosophy, we may, for a time, lose sight of the fact, that Practice without Science is Empiricism; and that all which is truly great in the practical must be the application and result of an antecedent ideal.

Its true nature.

What is the true system of education:

What, then, are the sources of that Utility, and the basis of that Practical, which the present generation desire, and after which they are so anxiously seeking? What system of training and discipline will best develop and steady the intellect of the young; give vigor and expansion to thought, and stability to action? What course of study will most enlarge the sphere of investigation; give the greatest freedom to the mind without licentiousness, and the greatest freedom to action consistent with the laws of nature, and the obligations of the social compact? What subject of study is, from its nature, most likely to ensure this training, and contribute to such results, and at the same time lay the foundations of all that is truly great in the Practical? It has seemed to me that mathematical science may lay claim to this pre-eminence.

Which will develop and steady the intellect?

What are the subjects of study?

Mathematics.

Foundations of mathematical knowledge. The first impressions which the child receives of Number and Quantity are the foundations of his mathematical knowledge. They form, as it were, a part of his intellectual being. Laws of Nature. The laws of Nature are merely truths or generalized facts, in regard to matter, derived by induction from experience, observation, and experiment. The laws of mathematical science are generalized Number and Space. truths derived from the consideration of Number and Space. All the processes of inquiry and investigation are conducted according to fixed laws, and form a science; and every new thought and higher impression form additional links in the lengthening chain.

Mathematical knowledge: The knowledge which mathematical science imparts to the mind is deep—profound—abiding. It gives rise to trains of thought, which are born in the pure ideal, and fed and nurtured by an acquaintance with physical nature in all its minuteness and in all its grandeur: What it does. which survey the laws of elementary organization, by the microscope, and weigh the spheres in the balance of universal gravitation.

What the processes effect. The processes of mathematical science serve to give mental unity and wholeness. They impart that knowledge which applies the means of crystallization to a chaos of scattered particulars, and discovers at once the general law, if there be one, which forms a connecting link between them. Such results can only be attained by minds highly disciplined by scientific combinations. In all these processes no fact of science is forgotten or lost. They are all engraved on the great tablet of universal truth, there to be read by succeeding generations so long as the laws of mind remain unchanged. This is strikingly illustrated by the fact, that any diligent student of a college may now read the works of Newton, or the Mécanique Céleste of La Place.

Right knowledge applies the means of crystallization.

It records and preserves truth.

The educator regards mathematical science as the great means of accomplishing his work. The definitions present clear and separate ideas, which the mind readily apprehends. The axioms are the simplest exercises of the reasoning faculty, and afford the most satisfactory results in the early use and employment of that faculty. The trains of reasoning which follow are combinations, according to logical rules, of what has been previously fully comprehended; and the mind and the argument grow together, so that the thread of science and the warp of the intellect entwine themselves, and become inseparable. Such a training will lay the foundations

How the educator regards mathematics.

The axioms.

Influence of the study of mathematics on the mind.

of systematic knowledge, so greatly preferable to conjectural judgments.

How the philosopher regards mathematics:

The philosopher regards mathematical science as the mere tools of his higher vocation. Looking with a steady and anxious eye to Nature, and the great laws which regulate and govern all things, he becomes earnestly intent on their examination, and absorbed in the wonderful harmonies which he discovers. Urged forward by these high impulses, he sometimes neglects that thorough preparation, in mathematical science, necessary to success; and is not unfrequently obliged, like Antæus, to touch again his mother earth, in order to renew his strength.

Its necessity to him.

The views of the practical man.

The mere practical man regards with favor only the results of science, deeming the reasonings through which these results are arrived at, quite superfluous. Such should remember that the mind requires instruments as well as the hands, and that it should be equally trained in their combinations and uses. Such is, indeed, now the complication of human affairs, that to do one thing well, it is necessary to know the properties and relations of many things. Every thing, whether existing in the abstract or in the material world; whether an element of knowl-

Instruments of the mind.

Every thing has a law.

edge or a rule of art, has its connections and its law: to understand these connections and that law, is to know the thing. When the principle is clearly apprehended, the practice is easy. To know the law is to know the thing.

With these general views, and under a firm conviction that mathematical science must become the great basis of education, I have bestowed much time and labor on its analysis, as a subject of knowledge. I have endeavored to present its elements separately, and in their connections; to point out and note the mental faculties which it calls into exercise; to show why and how it develops those faculties; and in what respect it gives to the whole mental machinery greater power and certainty of action than can be attained by other studies. To accomplish these ends, in the way that seemed to me most desirable, I have divided the work into three parts, arranged under the heads of Book I., II., and III. Mathematics analyzed. How. What was deemed necessary.

Book I. treats of Logic, both as a science and an art; that is, it explains the laws which govern the reasoning faculty, in the complicated processes of argumentation, and lays down the rules, deduced from those laws, for conducting such processes. It being one of the leading Logic. Explanation.

For what used. objects to show that mathematical science is the best subject for the development and application of the principles of logic; and, indeed, that the science itself is but the application of those principles to the abstract quantities Number and Space, it appeared indispensable to give, in a manner best adapted to my purpose, an outline of the nature of that reasoning by means of which all inferred knowledge is acquired.

The necessity of treating it.

Book II.

Book II. treats of Mathematical Science. Here I have endeavored to explain the nature of the subjects with which mathematical science is conversant; the ideas which arise in examining and contemplating those subjects; the language employed to express those ideas, and the laws of their connection. This, of course, led to a classification of the subjects; to an analysis of the language used, and an examination of the reasonings employed in the methods of proof.

Of what it treats.

Manner of treating.

Book III. Utility of Mathematics.

Book III. explains and illustrates the Utility of Mathematics: First, as a means of mental discipline and training; Secondly, as a means of acquiring knowledge; and, Thirdly, as furnishing those rules of art, which make knowledge practically effective.

Having thus given the general outlines of the work, we will refer to the classes of readers for whose use it is designed, and the particular advantages and benefits which each class may receive from its perusal and study. Classes of readers.

There are four classes of readers, who may, it is supposed, be profited, more or less, by the perusal of this work: Four classes

1st. The general reader; First class.
2d. Professional men and students; Second.
3d. Students of mathematics and philosophy; Third.
4th. Professional Teachers. Fourth.

First. The general reader, who reads for improvement, and desires to acquire knowledge, must carefully search out the import of language. He must early establish and carefully cultivate the habit of noting the connection between ideas and their signs, and also the relation of ideas to each other. Such analysis leads to attentive reading, to clear apprehension, deep reflection, and soon to generalization. Advantages to the general reader. Connection between words and ideas.

Logic considers the forms in which truth must be expressed, and lays down rules for reducing all trains of thought to such known forms. This habit of analyzing arms us with tests by which we separate argument from sophistry—truth from falsehood. The application of these principles, Logic. Its value:

In the study of mathematics.

in the construction of the mathematical science, where the relation between the sign (or language) and the thing signified (or idea expressed), is unmistakable, gives precision and accuracy, leads to right arrangement and classification, and thus prepares the mind for the reception of general knowledge.

Advantages to professional men.

Secondly. The increase of knowledge carries with it the necessity of classification. A limited number of isolated facts may be remembered, or a few simple principles applied, without tracing out their connections, or determining the places which they occupy in the science of general knowledge. But when these facts and principles are greatly multiplied, as they are in the learned professions; when the labors of preceding generations are to be examined, analyzed, compared; when new systems are to be formed, combining all that is valuable in the past with the stimulating elements of the present, there is occasion for the constant exercise of our highest faculties. Knowledge reduced to order; that is, knowledge so classified and arranged as to be easily remembered, readily referred to, and advantageously applied, will alone suffice to sift the pure metal from the dust of ages, and fashion it for present use. Such knowledge is Science.

The reason.

Knowledge reduced to order is science.

Masses of facts, like masses of matter, are capable of very minute subdivisions; and when we know the law of combination, they are readily divided or reunited. To know the law, in any case, is to ascend to the source; and without that knowledge the mind gropes in darkness.

Knowledge may be reduced to its elements.

It has been my aim to present such a view of Logic and Mathematical Science as would clearly indicate, to the professional student, and even to the general reader, the outlines of these subjects. Logic exhibits the general formula applicable to all kinds of argumentation, and mathematics is an application of logic to the abstract quantities Number and Space.

Objects of the work.

Logic and mathematics.

When the professional student shall have examined the subject, even to the extent to which it is here treated, he will be impressed with the clearness, simplicity, certainty, and generality of its principles; and will find no difficulty in making them available in classifying the facts, and examining the organic laws which characterize his particular department of knowledge.

Certainty of the results.

Thirdly. Mathematical knowledge differs from every other kind of knowledge in this: it is, as it were, a web of connected principles spun out from a few abstract ideas, until it has become one of the great means of intellectual develop-

Mathematical knowledge.

Its extent.

ment and of practical utility. And if I am permitted to extend the figure, I may add, that the web of the spider, though perfectly simple, if we see the end and understand the way in which it is put together, is yet too complicated to be unravelled, unless we begin at the right point, and *observe the law of its formation.* So with mathematical science. It is evolved from a few —a very few—elementary and intuitive principles: the law of its evolution is simple but exacting, and to begin at the right place and proceed in the right way, is all that is necessary to make the subject easy, interesting, and useful.

Necessity of beginning at the right place.

How mathematical science is constructed.

I have endeavored to point out the place of beginning, and to indicate the way to the mathematical student. I am aware that he is starting on a road where the guide-boards resemble each other, and where, for the want of careful observation, they are often mistaken; I have sought, therefore, to furnish him with the maps and guide-books of an old traveller.

What has been attempted.

By explaining with minuteness the subjects about which mathematical science is conversant, the whole field to be gone over is at once surveyed: by calling attention to the faculties of the mind which the science brings into exercise, we are better prepared to note the intellectual operations which the processes require; and by

Advantages of examining the whole subject.

Advantages of considering the mental faculties:

a knowledge of the laws of reasoning, and an acquaintance with the tests of truth, we are enabled to verify all our results. These means have been furnished in the following work, and to aid the student in classification and arrangement, diagrams have been prepared exhibiting separately and in connection all the principal parts of mathematical science. The student, therefore, who adopts the system here indicated, will find his way clearly marked out, and will recognise, from their general resemblance to the descriptions, all the guide-posts which he meets. He will be at no loss to discover the connection between the parts of his subject. Beginning with first principles and elementary combinations, and guided by simple laws, he will go forward from the exercises of Mental Arithmetic to the higher analysis of Mathematical Science on an ascent so gentle, and with a progress so steady, as scarcely to note the changes. And indeed, why should he? For all mathematical processes are alike in their nature, governed by the same laws, exercising the same faculties, and lifting the mind towards the same eminence.

Of a knowledge of the laws of reasoning.

What has been done.

Advantages to the student.

Where he begins.

Order of progress.

Unity of the subject.

Fourthly. The leading idea, in the construction of the work, has been, to afford substantial aid to the professional teacher. The nature of

Advantages to the professional teacher.

His duties: his duties—their inherent difficulties—the perplexities which meet him at every step—the want of sympathy and support in his hours of discouragement — (and they are many) — are circumstances which awaken a lively interest in the hearts of all who have shared the toils, and been themselves laborers in the same vineyard. He takes his place in the schoolhouse by the roadside, and there, removed from the highways of life, spends his days in raising the feeble mind of childhood to strength—in planting aright the seeds of knowledge—in curbing the turbulence of passion — in eradicating evil and inspiring good. The fruits of his labors are seen but once in a generation. The boy must grow to manhood and the girl become a matron before he is certain that his labors have not been in vain.

Discouragements and difficulties:

Remoteness from active life.

Fruits of his efforts, when seen.

Yet, to the teacher is committed the high trust of forming the intellectual, and, to a certain extent, the moral development of a people. He holds in his hands the keys of knowledge. If the first moral impressions do not spring into life at his bidding, he is at the source of the stream, and gives direction to the current. Although himself imprisoned in the schoolhouse, his influence and his teachings affect all conditions of society, and reach over the whole hori-

The importance of his labors.

zon of civilization. He impresses himself on the young of the age in which he lives, and lives again in the age which succeeds him. The influence of his labors.

All good teaching must flow from copious knowledge. The shallow fountain cannot emit a vigorous stream. In the hope of doing something that may be useful to the professional teacher, I have attempted a careful and full analysis of mathematical science. I have spread out, in detail, those methods which have been carefully examined and subjected to the test of long experience. If they are the right methods, they will serve as standards of teaching; for, the principles of imparting instruction are the same for all branches of knowledge. Sources of good teaching. Objects for which the work was undertaken. Principles of all teaching, the same.

The system which I have indicated is complete in itself. It lays open to the teacher the entire skeleton of the science—exhibits all its parts separately and in their connection. It explains a course of reasoning simple in itself, and applicable not only to every process in mathematical science, but to all processes of argumentation in every subject of knowledge. System. What it presents. What it explains.

The teacher who thus combines science with art, no longer regards Arithmetic as a mere treadmill of mechanical labor, but as a means— Science combined with art:

The advantages resulting from it.

Results of right instruction.

and the simplest means—of teaching the art and science of reasoning on quantity—and this is the logic of mathematics. If he would accomplish well his work, he must so instruct his pupils that they shall apprehend clearly, think quickly and correctly, reason justly, and open their minds freely to the reception of all knowledge.

# BOOK I.

# LOGIC.

## CHAPTER I.

DEFINITIONS—OPERATIONS OF THE MIND—TERMS DEFINED.

### DEFINITIONS.

§ 1. DEFINITION is a metaphorical word, which literally signifies "laying down a boundary." All definitions are of names, and of names only; but in some definitions, it is clearly apparent, that nothing is intended except to explain the meaning of the word; while in others, besides explaining the meaning of the word, it is also implied that there exists, or may exist, a *thing* corresponding to the word.

Definition a metaphorical word.

Some definitions explain only words: others imply *things* corresponding to the words.

§ 2. Definitions which do not imply the existence of things corresponding to the words defined, are those usually found in the Dictionary of one's own language. They explain only the

Of definitions which do not imply things corresponding to words.

They explain words by equivalents.

meaning of the word or term, by giving some equivalent expression which may happen to be better known. Definitions which imply the existence of things corresponding to the words defined, do more than this.

Definition of a triangle; what it implies.

For example: "A triangle is a rectilineal figure having three sides." This definition does two things:

1st. It explains the meaning of the word triangle; and,

2d. It implies that there exists, or may exist, a rectilineal figure having three sides.

Of a definition which implies the existence of a thing.

§ 3. To define a word when the definition is to imply the existence of a thing, is to select from all the properties of the thing those which are most simple, general, and obvious; and the properties must be very well known to us before we can decide which are the fittest for this purpose. Hence, a thing may have many properties besides those which are named in the definition of the word which stands for it. This second kind of definition is not only the best form of expressing certain conceptions, but also contributes to the development and support of new truths.

Properties must be known.

A definition supports truth.

In Mathematics names imply

§ 4. In Mathematics, and indeed, in all strict sciences, names imply the existence of the things which they name; and the definitions of those names express attributes of the things; so that no correct definition whatever, of any mathematical term, can be devised, which shall not express certain attributes of the thing corresponding to the name. Every definition of this class is a tacit assumption of some proposition which is expressed by means of the definition, and which gives to such definition its importance.

things and express attributes.

Definitions of this class are propositions.

§ 5. All the reasonings in mathematics, which rest ultimately on definitions, do, in fact, rest on the intuitive inference, that things corresponding to the words defined have a *conceivable* existence as subjects of thought, and do or may have *proximately*, an *actual* existence.*

Reasoning resting on definitions;

rests on intuitive inferences.

---

* There are four rules which aid us in framing definitions.

Four rules.

1st. The definition must be *adequate:* that is, neither too extended, nor too narrow for the word defined.

1st rule.

2d. The definition must be in itself *plainer* than the word defined, else it would not explain it.

2d rule.

3d. The definition should be expressed in a *convenient number of appropriate words.*

3d rule.

4th. When the definition implies the existence of a thing corresponding to the word defined, the certainty of that existence must be intuitive.

4th rule.

OPERATIONS OF THE MIND CONCERNED IN REASONING.

Three operations of the mind.

§ 6. There are three operations of the mind which are immediately concerned in reasoning.

1st. Simple apprehension; 2d. Judgment; 3d. Reasoning or Discourse.

Simple apprehension.

§ 7. Simple apprehension is the notion (or conception) of an object in the *mind*, analogous to the perception of the senses. It is either Incomplex or Complex. Incomplex Apprehension is of one object, or of several without any *relation* being perceived between them, as of a triangle, a square, or a circle: Complex is of several with such a *relation*, as of a triangle within a circle, or a circle within a square.

Incomplex.

Complex.

Judgment defined.

§ 8. Judgment is the comparing together in the mind two of the notions (or ideas) which are the objects of apprehension, whether complex or incomplex, and pronouncing that they agree or disagree with each other, or that one of them *belongs* or does *not* belong to, the other: for example: that a right-angled triangle and an equilateral triangle belong to the class of figures called triangles; or that a square is *not* a circle. Judgment, therefore, is either *Affirmative* or *Negative*.

Judgment is either affirmative or negative.

§ 9. Reasoning (or discourse) is the act of proceeding from certain judgments to another *founded* upon them (or the result of them).

Reasoning defined.

§ 10. Language affords the *signs* by which these operations of the mind are recorded, expressed, and communicated. It is also an instrument of thought, and one of the principal helps in all mental operations; and any imperfection in the instrument, or in the mode of using it, will materially affect any result attained through its aid.

Language affords signs of thought: also, an instrument of thought.

§ 11. Every branch of knowledge has, to a certain extent, its own appropriate language; and for a mind not previously versed in the meaning and right use of the various words and signs which constitute the language, to attempt the study of methods of philosophizing, would be as absurd as to attempt reading before learning the alphabet.

Every branch of knowledge has its own language, which must be learned.

## ABSTRACTION.

§ 12. The faculty of abstraction is that power of the mind which enables us, in contemplating any object (or objects), to attend exclusively to

Abstraction,

some particular circumstance belonging to it, and quite withhold our attention from the rest. Thus, if a person in contemplating a rose should make the scent a distinct object of attention, and lay aside all thought of the form, color, &c., he would *draw off*, or *abstract* that particular part; and therefore employ the faculty of *abstraction.* He would also employ the same faculty in considering whiteness, softness, virtue, existence, as entirely separate from particular objects.

in contemplating a rose: the process of drawing off.

§ 13. The term *abstraction*, is also used to denote the *operation* of abstracting from one or more things the particular part under consideration; and likewise to designate the *state* of the mind when occupied by abstract ideas. Hence, abstraction is used in three senses:

The term Abstraction, how used.

1st. To denote a faculty or power of the mind;

2d. To denote a process of the mind; and,

3d. To denote a state of the mind.

Abstraction denotes a faculty, a process, and a state of mind.

GENERALIZATION.

§ 14. Generalization is the process of contemplating the agreement of several objects in certain points (that is, abstracting the circumstances of agreement, disregarding the differ-

Generalization—the process of contemplating the agreement.

ences), and giving to all and each of these objects a name applicable to them in respect to this agreement. For example; we give the name of triangle, to every rectilineal figure having *three* sides: thus we *abstract* this property from all the others (for, the triangle has three angles, may be equilateral, or scalene, or right-angled), and name the entire class from the property so abstracted. Generalization therefore necessarily implies abstraction; though abstraction does not imply generalization.

of several things.

Generalization

implies abstraction.

TERMS—SINGULAR TERMS—COMMON TERMS.

§ 15. An act of apprehension, expressed in language, is called a Term. Proper names, or any other terms which denote each but a single individual, as "Cæsar," "the Hudson," "the Conqueror of Pompey," are called Singular Terms.

A term.

Singular terms.

On the other hand, those terms which denote any individual of a whole class (which are formed by the process of abstraction and generalization), are called Common or general Terms. For example; quadrilateral is a common term, applicable to every rectilineal plane figure having four sides; River, to all rivers; and Conqueror, to all conquerors. The individuals for which a common term stands, are called its *Significates.*

Common terms.

Significates.

## CLASSIFICATION.

Classification.

§ 16. Common terms afford the means of classification; that is, of the arrangement of objects into classes, with reference to some common and distinguishing characteristic. A collection, comprehending a number of objects, so arranged, is called a Genus or Species—genus being the more extensive term, and often embracing many species.

Genus, species.

Examples in classification.

For example: animal is a genus embracing every thing which is endowed with life, the power of voluntary motion, and sensation. It has many species, such as man, beast, bird, &c. If we say of an animal, that it is *rational*, it belongs to the species man, for this is the characteristic of that species. If we say that it has wings, it belongs to the species bird, for this, in like manner, is the characteristic of the species bird.

Subspecies or classes.

A species may likewise be divided into classes, or subspecies; thus the species man, may be divided into the classes, male and female, and these classes may be again divided until we reach the individuals.

Principles of classification.

§ 17. Now, it will appear from the principles which govern this system of classification, that the characteristic of a genus is of a more extensive signification, but involves fewer particulars than that of a species. In like manner, the characteristic of a species is more extensive, but *less full* and *complete*, than that of a subspecies or class, and the characteristics of these less full than that of an individual.

Genus more extensive than species,

but less full and complete.

For example; if we take as a genus the Quadrilaterals of Geometry, of which the characteristic is, that they have four sides, then every plane rectilineal figure, having four sides, will fall under this class. If, then, we divide all quadrilaterals into two species, viz. those whose opposite sides, taken two and two, are not parallel, and those whose opposite sides, taken two and two, are parallel, we shall have in the first class, all irregular quadrilaterals, including the trapezoid (1 and 2); and in the other, the parallelogram, the rhombus, the rectangle, and the square (3, 4, 5, and 6).

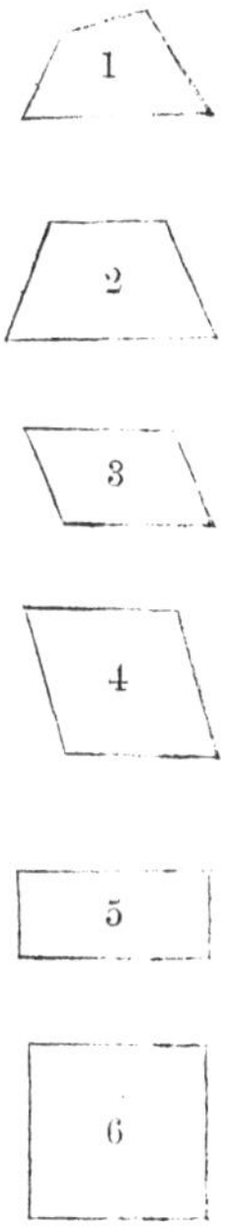

If, then, we divide the first species into two subspecies or classes, we shall have in the one, the irregular quadrilaterals (1), and in the other, the trapezoids (2); and each of these classes, being made up of individuals having the *same characteristics*, are not susceptible of further division.

If we divide the second species into two classes, arranging those which have oblique angles in the one, and those which have right angles in the other, we shall have in the first, two varieties, viz. the common parallelogram and the equilateral parallelogram or rhombus (3 and 4); and in the second, two varieties also, viz. the rectangle and the square (5 and 6).

Species and classes.

Now, each of these six figures is a *quadrilateral;* and hence, possesses the *characteristic* of the genus; and each variety of both species enjoys all the characteristics of the species to which it belongs, together with some other distinguishing feature; and similarly, of *all classifications.*

Each individual falling under the genus enjoys all the characteristics.

§ 18. In special classifications, it is often not necessary to begin with the most general characteristics; and then the genus with which we begin, is in fact but a *species of a more extended classification,* and is called a Subaltern Genus.

Subaltern genus.

For example; if we begin with the genus Parallelogram, we shall at once have two species, viz. those parallelograms whose angles are oblique and those whose angles are right angles; and in each species there will be two varieties, viz. in the first, the common parallelogram and the rhombus; and in the second, the rectangle and square.

Parallelogram.

§ 19. A genus which cannot be considered as a species, that is, which cannot be referred

Highest genus.

to a more extended classification, is called the highest genus; and a species which cannot be considered as a genus, because it contains only individuals having the same characteristic, is called the lowest species.

Highest genus.

Lowest species.

NATURE OF COMMON TERMS.

§ 20. It should be steadily kept in mind, that the "common terms" employed in classification, have not, as the names of individuals have, any *real existing thing in nature* corresponding to them; but that each is merely a name denoting a certain *inadequate notion* which our minds have formed of an *individual*. But as this name does not include any thing wherein that individual differs from others of the same class, it is applicable equally well to all or any of them. Thus, quadrilateral denotes no *real thing*, distinct from each individual, but merely *any* rectilineal figure of four sides, viewed *inadequately;* that is, after *abstracting* and *omitting* all that is *peculiar* to each individual of the class. By this means, a common term becomes applicable alike to any one of several individuals, or, taken in the plural, to several individuals together.

A common term has no real thing correspond-ing:

is an inadequate notion:

does not include any thing in which individuals differ;

but is applicable to many individuals.

Much needless difficulty has been raised respecting the results of this process: many having contended, and perhaps more having taken

Needless difficulty.

Difficulty in the interpretation of common terms.

it for granted, that there must be some really existing *thing* corresponding to each of those common terms, and of which such term is the name, standing for and representing it. For example; since there is a really existing thing corresponding to and signified by the proper and singular name "Ætna," it has been supposed that the *common* term "Mountain" must have some one really existing thing corresponding to it, and of course *distinct* from each individual mountain, yet existing *in* each, since the term, being common, is applicable, separately, to every one of them.

No one real thing corresponding to each.

The fact is, the notion expressed by a common term is merely an inadequate (or incomplete) notion of an individual; and from the very circumstance of its inadequacy, it will apply equally well to any one of several individuals. For example; if I omit the mention and the consideration of every circumstance which distinguishes Ætna from any other mountain, I then form a notion, that inadequately designates Ætna. This notion is expressed by the common term "mountain," which does not imply any of the peculiarities of the mountain Ætna, and is equally applicable to any one of several individuals.

Merely an inadequate notion partially designating the thing.

"Mountain" is applicable to all mountains.

In regard to classification, we should also bear in mind, that we may fix, arbitrarily, on the characteristic which we choose to abstract and consider as the basis of our classification, disregarding all the rest: so that the same individual may be referred to any of several different species, and the same species to several genera, as suits our purpose.

May fix on attributes arbitrarily for classification.

### SCIENCE.

§ 21. Science, in its popular signification, means knowledge.* In a more restricted sense, it means knowledge reduced to order; that is, knowledge so classified and arranged as to be easily remembered, readily referred to, and advantageously applied. In a more strict and technical sense, it has another signification.

Science in its general sense.

Has a technical signification.

"Every thing in nature, as well in the inanimate as in the animated world, happens or is done according to rules, though we do not always know them. Water falls according to the laws of gravitation, and the motion of walking is performed by animals according to rules. The fish in the water, the bird in the air, move according to rules. There is nowhere any want of rule. When we think we find that want, we can only say that, in this case, the rules are unknown to us."†

Views of Kant.

General laws.

Nowhere any want of rule.

Assuming that all the phenomena of nature

* Section 23. † Kant.

Science in a technical sense defined: is an analysis of the laws of nature.

are consequences of general and immutable laws, we may define *Science* to be the analysis of those laws,—comprehending not only the connected processes of experiment and reasoning which make them known to man, but also those processes of reasoning which make known their individual and concurrent operation in the development of individual phenomena.

### ART.

Art, application of science,

§ 22. Art is the application of knowledge to practice. Science is conversant about knowledge: Art is the use or application of knowledge, and is conversant about works. Science has knowledge for its object: Art has knowledge for its guide. A principle of science, when applied, becomes a rule of art. The developments of science increase knowledge: the applications of art add to works. Art, necessarily, presupposes knowledge: art, in any but its infant state, presupposes scientific knowledge; and if every art does not bear the name of the science on which it rests, it is only because several sciences are often necessary to form the groundwork of a single art. Such is the complication of human affairs, that to enable one thing to be *done*, it is often requisite to *know* the nature and properties of many things.

and presupposes knowledge.

Many things must be known before one can be done.

# CHAPTER II.

## SOURCES AND MEANS OF KNOWLEDGE—INDUCTION.

### KNOWLEGDE.

§ 23. KNOWLEDGE is a clear and certain conception of that which is true, and implies three things:

Knowledge a clear conception of what is true:

1st. Firm belief; 2d. Of what is true; and, 3d. On sufficient grounds.

Implies—1st. Firm belief; 2d. Of what is true; 3d. On sufficient grounds.

If any one, for example, is in *doubt* respecting one of Legendre's Demonstrations, he cannot be said to *know* the proposition proved by it. If, again, he is fully *convinced* of any thing that is not *true*, he is mistaken in supposing himself to *know* it; and lastly, if two persons are each *fully confident*, one that the moon is inhabited, and the other that it is not (though one of these opinions must be *true*), neither of them could properly be said to *know* the truth, since he cannot have sufficient *proof* of it.

Examples.

### FACTS AND TRUTHS.

Knowledge is of facts and truths.

§ 24. Our knowledge is of two kinds: of facts and truths. A fact is any thing that HAS BEEN or IS. That the sun rose yesterday, is a fact: that he gives light to-day, is a fact. That water is fluid and stone solid, are facts. We derive our knowledge of facts through the medium of the senses.

Truth an accordance with what has been, is, or shall be.

Truth is an exact accordance with what HAS BEEN, IS, OR SHALL BE. There are two methods of ascertaining truth:

Two methods of ascertaining it.

1st. By comparing known facts with each other; and,

2dly. By comparing known truths with each other.

Hence, truths are inferences either from facts or other truths, made by a mental process called Reasoning.

Facts and truths, the elements of our knowledge.

§ 25. Seeing, then, that facts and truths are the elements of all our knowledge, and that knowledge itself is but their clear apprehension, their firm belief, and a distinct conception of their relations to each other, our main inquiry is, How are we to attain unto these facts and truths, which are the foundations of knowledge?

1st. Our knowledge of facts is derived through

the medium of our senses, by observation, experiment,* and experience. We see the tree, and perceive that it is shaken by the wind, and note the fact that it is in motion. We decompose water and find its elements; and hence, learn from experiment the *fact*, that it is not a simple substance. We experience the vicissitudes of heat and cold; and thus learn from experience that the temperature is not uniform.

How we arrive at a knowledge of facts.

The ascertainment of facts, in any of the ways above indicated, does not point out any connection between them. It merely exhibits them to the mind as separate or isolated; that is, each as standing for a determinate thing, whether simple or compound. The term facts, in the sense in which we shall use it, will designate facts of this class only. If the facts so ascertained have such connections with each other, that additional facts can be inferred from them, that inference is pointed out by the reasoning process, which is carried on, in all cases, by comparison.

This does not point out a connection between them.

When they have a connection that is pointed out by the reasoning process.

2dly. A result obtained by comparing facts, we have designated by the term Truth. Truths, therefore, are inferences from facts; and every

Truth, found by comparing facts;

* Under this term we include all the methods of investigation and processes of arriving at facts, except the process of reasoning.

and is inferred from them.

truth has reference to all the singular facts from which it is inferred. Truths, therefore, are results deduced from facts, or from classes of facts. Such results, when obtained, appertain to all facts of the same class. Facts make a genus: truths, a species; with the characteristic, that they become known to us by inference or reasoning.

How truths are inferred from facts by the reasoning process.

§ 26. How, then, are truths to be inferred from facts by the reasoning process? There are two cases.

1st case.

1st. When the instances are so few and simple that the mind can contemplate *all the facts* on which the induction rests, and to which it refers, and can make the induction without the aid of other facts; and,

2d case.

2dly. When the facts, being numerous, complicated, and remote, are brought to mind only by processes of investigation.

## INTUITIVE TRUTH.

Intuitive or Self-evident truths.

§ 27. Truths which become known by considering all the facts on which they depend, and which are inferred the moment the facts are apprehended, are the subjects of Intuition, and are called *Intuitive* or *Self-evident Truths.* The

Intuition defined.

term Intuition is strictly applicable only to that mode of contemplation in which we look at facts, or classes of facts, and apprehend the relations of those facts at the same time, and by the same act by which we apprehend the facts themselves. Hence, intuitive or self-evident truths are those which are conceived in the mind immediately; that is, which are perfectly conceived by a single process of induction, the moment the facts on which they depend are apprehended, without the intervention of other ideas. They are necessary consequences of conceptions respecting which they are asserted. The axioms of Geometry afford the simplest and most unmistakable class of such truths.

How intuitive truths are conceived in the mind.

Axioms of Geometry are the simplest kind.

"A whole is equal to the sum of all its parts," is an intuitive or self-evident truth, inferred from facts previously learned. For example; having learned from experience and through the senses what a whole is, and, from experiment, the fact that it may be divided into parts, the mind perceives the relation between the whole and the sum of the parts, viz. that they are equal; and then, by the reasoning process, infers that the same will be true of every other thing; and hence, pronounces the general truth, that "a whole is equal to the sum of all its parts." Here all the facts from which the induction is drawn, are presented to the mind, and the induction is made without the aid of other facts; hence,

A whole equal to the sum of all the parts, an intuitive truth.

How inferred.

All the *facts* are presented to the mind.

All the axioms are deduced in the same way.

it is an intuitive or self-evident truth. All the other axioms of Geometry are deduced from premises and by processes of inference, entirely similar. We would not call these experimental truths, for they are not alone the results of experiment or experience. Experience and experiment furnish the *requisite information,* but the *reasoning power* evolves the general truth.

These axioms are general propositions.

"When we say, the equals of equals are equal, we mentally make comparisons in equal spaces, equal times, &c.; so that these axioms, however self-evident, are still general propositions: so far of the inductive kind, that, independently of experience, they would not present themselves to the mind. The only difference between these and axioms obtained from extensive induction is this: that, in raising the axioms of Geometry, the instances offer themselves spontaneously, and without the trouble of search, and are few and simple: in raising those of nature, they are infinitely numerous, complicated, and remote; so that the most diligent research and the utmost acuteness are required to unravel their web, and place their meaning in evidence."*

Difference between them and other propositions, which require diligent research.

* Sir John Herschel's Discourse on the study of Natural Philosophy.

## TRUTHS, OR LOGICAL TRUTHS.

Truths generalized from facts, and truths inferred from truths.

§ 28. Truths inferred from facts, by the process of generalization, when the instances do not offer themselves spontaneously to the mind, but require search and acuteness to discover and point out their connections, and all truths inferred from truths, might be called *Logical Truths.* But as we have given the name of intuitive or self-evident truths to all inferences in which *all the facts* were contemplated, we shall designate all others by the simple term, TRUTHS.

Necessity of the distinction, being the basis of a classification.

It might appear of little consequence to distinguish the processes of reasoning by which truths are inferred from facts, from those in which we deduce truths from other truths; but this difference in the premises, though seemingly slight, is nevertheless very important, and divides the subject of logic, as we shall presently see, into two distinct and very different branches.

## LOGIC.

Logic notes the sufficiency of evidence.

§ 29. Logic takes note of and decides upon the sufficiency of the evidence by which truths are established. Our assent to the conclusion being grounded on the truth of the premises, we never could arrive at any knowledge by reasoning, unless something were known antecedently to all reasoning. It is the province of

Its province.

Furnishes the tests of truth.

Logic to furnish the tests by which all truths that are not intuitive may be inferred from the premises. It has nothing to do with ascertaining facts, nor with any proposition which claims to be believed on its own intrinsic evidence; that is, without evidence, in the proper sense of the word. It has nothing to do with the original data, or ultimate premises of our knowledge; with their number or nature, the mode in which they are obtained, or the tests by which they are distinguished. But, so far as our knowledge is founded on truths made such by evidence, that is, derived from facts or other truths previously known, whether those truths be particular truths, or general propositions, it is the province of Logic to supply the tests for ascertaining the validity of such evidence, and whether or not a belief founded on it would be well grounded. And since by far the greatest portion of our knowledge, whether of particular or general truths, is avowedly matter of inference, nearly the whole, not only of science, but of human conduct, is amenable to the authority of logic.

Has nothing to do with intuitive propositions, nor with original data;

but supplies all tests for general propositions.

The greatest portion of our knowledge comes from inference.

## INDUCTION.

§ 30. That part of logic which infers truths from facts, is called Induction. Inductive reasoning is the application of the reasoning process to a given number of facts, for the purpose of determining if what has been ascertained respecting one or more of the individuals is true of the whole class. Hence, Induction is not the mere sum of the facts, but a conclusion drawn from them.

Induction, to what reasoning applicable.

Induction defined.

The logic of Induction consists in classing the facts and stating the inference in such a manner, that the evidence of the inference shall be most manifest.

Logic of Induction.

§ 31. Induction, as above defined, is a process of inference. It proceeds from the known to the unknown; and any operation involving no inference, any process in which the conclusion is a mere fact, and not a truth, does not fall within the meaning of the term. The conclusion must be broader than the premises. The premises are facts: the conclusion must be a truth.

Induction proceeds from the known to the unknown.

The conclusion broader than the premises.

Induction, therefore, is a process of generalization. It is that operation of the mind by which we infer that what we know to be true

Induction, a process of generalization;

in which we conclude, that what is true under particular circumstances will be true universally.

in a particular case or cases, will be true in all cases which resemble the former in certain assignable respects. In other words, Induction is the process by which we conclude that what is true of certain individuals of a class is true of the whole class; or that what is true at certain times, will be true, under similar circumstances, at all times.

Induction presupposes accurate and necessary observations.

§ 32. Induction always presupposes, not only that the necessary observations are made with the necessary accuracy, but also that the results of these observations are, so far as practicable, connected together by general descriptions: enabling the mind to represent to itself as wholes, whatever phenomena are capable of being so represented.

More is necessary than to connect the observations: we must *infer* from them.

To suppose, however, that nothing more is required from the conception than that it should serve to connect the observations, would be to substitute hypothesis for theory, and imagination for proof. The connecting link must be some character which *really exists* in the facts themselves, and which would manifest itself therein, if the condition could be realized which our organs of sense require.

For example; Blakewell, a celebrated English cattle-breeder, observed, in a great number of individual beasts, a tendency to fatten readily, and in a great number of others the absence of this constitution: in every individual of the former description, he *observed* a certain peculiar *make*, though they differed widely in size, color, &c. Those of the latter description differed no less in various points, but agreed in being of a different make from the others. These *facts* were his data; from which, combining them with the general principle, that nature is steady and uniform in her proceedings, he *logically* drew the conclusion that beasts of the specified make have *universally* a peculiar tendency to fattening.

Example of Blakewell, the English cattle breeder.

How he ascertained the facts: why he inferred.

The principal difficulty in this case consisted in *making* the observations, and so *collating* and *combining* them as to *abstract* from each of a multitude of cases, differing widely in many respects, the circumstances in which they all agreed. But neither the making of the observations, nor their combination, nor the abstraction, nor the judgment employed in these processes, constituted the induction, though they were all preparatory to it. The Induction consisted in the generalization; that is, in *inferring* from all the data, that certain circumstances would be found in the whole class.

In what the difficulty consisted.

In what the induction consisted.

The mind of Newton was led to the universal law, that all bodies attract each other by forces

Newton's inference of the law of universal gravitation.

varying directly as their masses, and inversely as the squares of their distances, by Induction. He saw an apple falling from the tree: a mere fact; and asked himself the cause; that is, if any *inference* could be drawn from that fact, which should point out an invariable antecedent condition. This led him to note other facts, to prosecute experiments, to observe the heavenly bodies, until from many facts, and *their connections with each other,* he arrived at the conclusion, that the motions of the heavenly bodies were governed by general laws, applicable to all matter; that the stone whirled in the sling and the earth rolling forward through space, are governed in their motions by one and the same law. He then brought the exact sciences to his aid, and demonstrated that this law accounted for all the phenomena, and harmonized the results of all observations. Thus, it was ascertained that the laws which regulate the motions of the heavenly bodies, as they circle the heavens, also guide the feather, as it is wafted along on the passing breeze.

How he observed facts and their connections.

The use which he made of exact science.

What was the result.

The ways of ascertaining facts are known:

§ 33. We have already indicated the ways in which the facts are ascertained from which the inferences are drawn. But when an inference can be drawn; how many facts must enter into the premises; what their exact nature must be; and what their relations to each other, and to the inferences which flow from them; are questions which do not admit of definite answers. Although no general law has yet been discovered connecting all facts with truths, yet all the uniformities which exist in the succession of phenomena, and most of those which prevail in their coexistence, are either themselves laws of causation or consequences resulting and corollaries capable of being deduced from, such laws. It being the main business of Induction to determine the effects of every cause, and the causes of all effects, if we had for all such processes general and certain laws, we could determine, in all cases, what causes are correctly assigned to what effects, and what effects to what causes, and we should thus be virtually acquainted with the whole course of nature. So far, then, as we can trace, *with certainty*, the connection between cause and effect, or between effects and their causes, to that extent Induction is a science. When this cannot be done, the conclusions must be, to some extent, conjectural.

but we do not know certainly, in all cases, when we can draw on inference.

No general law.

Business of Induction.

What is necessary.

How far a science.

## CHAPTER III.

### DEDUCTION—NATURE OF THE SYLLOGISM—ITS USES AND APPLICATIONS.

#### DEDUCTION.

Inductive processes of reasoning.

Deductive processes.

Deduction defined.

Deductive formula.

§ 34. We have seen that all processes of Reasoning, in which the premises are particular facts, and the conclusions general truths, are called Inductions. All processes of Reasoning, in which the premises are general truths and the conclusions particular truths, are called Deductions. Hence, a deduction is the process of reasoning by which a particular truth is inferred from other truths which are known or admitted. The formula for all deductions is found in the Syllogism, the parts, nature, and uses of which we shall now proceed to explain.

#### PROPOSITIONS.

Proposition, judgment in words:

§ 35. A proposition is a *judgment expressed in words.* Hence, a proposition is defined logically, "A sentence indicative:" affirming or

* Section 30.

denying; therefore, it must not be ambiguous, for that which has more than one meaning is in reality several propositions; nor *imperfect*, nor *ungrammatical*, for such expressions have no meaning at all.

must not be ambiguous; nor imperfect; nor ungrammatical.

§ 36. Whatever can be an object of belief, or even of disbelief, must, when put into words, assume the form of a proposition. All truth and all error lie in propositions. What we call a truth, is simply a true proposition; and errors are false propositions. To know the import of all propositions, would be to know all questions which can be raised, and all matters which are susceptible of being either believed or disbelieved. Since, then, the objects of all belief and all inquiry express themselves in propositions, a sufficient scrutiny of propositions and their varieties will apprize us of what questions mankind have actually asked themselves, and what, in the nature of answers to those questions, they have actually thought they had grounds to believe.

A proposition explained.

Its nature,—extent.

Embraces all truth and all error.

An examination of propositions embraces all questions and all knowledge.

§ 37. The first glance at a proposition shows that it is formed by putting together two names. Thus, in the proposition, "Gold is yellow," the property *yellow* is affirmed of the substance *gold*. In the proposition, "Franklin was not born in

A proposition is formed by putting two names together.

England," the fact expressed by the words *born in England* is denied of the man Franklin.

A proposition has three parts: Subject, Predicate, and Copula.

§ 38. Every proposition consists of three parts: the Subject, the Predicate, and the Copula. The subject is the name denoting the person or thing of which something is affirmed or denied: the predicate is that which is affirmed or denied of the subject; and these two are called the *terms* (or extremes), because, logically, the subject is placed *first,* and the predicate *last.* The copula, in the middle, indicates the act of judgment, and is the sign denoting that there is an affirmation or denial. Thus, in the proposition, "The earth is round;" the subject is the words "the earth," being that of which something is affirmed: the predicate, is the word *round,* which denotes the quality affirmed, or (as the phrase is) *predicated:* the word IS, which serves as a connecting mark between the subject and the predicate, to show that one of them is affirmed of the other, is called the Copula. The copula must be either IS, or IS NOT, the substantive verb being the only *verb* recognised by Logic. All other verbs are resolvable, by means of the verb "to be," and a participle or adjective. For example:

Subject defined.

Predicate.

Copula must be IS OR IS NOT.

All verbs resolvable into "to be."

"The Romans conquered:"

the word "*conquered*" is both copula and predicate, being equivalent to "*were victorious.*"

Examples of the Copula.

Hence, we might write,

"The Romans were victorious,"

in which *were* is the copula, and *victorious* the predicate.

§ 39. A proposition being a portion of discourse, in which something is affirmed or denied of something, all propositions may be divided into affirmative and negative. An affirmative proposition is that in which the predicate is affirmed of the subject; as, "Cæsar is dead." A negative proposition is that in which the predicate is *denied* of the subject; as, "Cæsar is not dead." The copula, in this last species of proposition, consists of the words "IS NOT," which is the sign of negation; "IS" being the sign of affirmation.

A proposition is either affirmative or negative

In the last, the copula is, IS NOT.

SYLLOGISM.

§ 40. A syllogism is a form of stating the connection which may exist, for the purpose of reasoning, between three propositions. Hence, to a legitimate syllogism, it is essential that there should be three, and only three, proposi-

A syllogism consists of three propositions.

Two are admitted;

and the third is proved from them.

tions. Of these, two are admitted to be true, and are called the *premises:* the third is proved from these two, and is called the conclusion. For example:

Example.

> "All tyrants are detestable:
> Cæsar was a tyrant;
> *Therefore*, Cæsar was detestable."

Now, if the first two propositions be *admitted,* the third, or conclusion, necessarily follows from them, and it is proved that CÆSAR was detestable.

Major Term defined.

Minor Term.

Of the two terms of the conclusion, the Predicate (detestable) is called the *major term,* and the Subject (Cæsar) the *minor term;* and these two terms, together with the term "tyrant," make up the three propositions of the syllogism, —each term being used twice. Hence, every syllogism has three, and only three, *different* terms.

Major Premiss defined.

Minor Premiss.

Middle Term.

The premiss, into which the Predicate of the conclusion enters, is called the *major premiss;* the other is called the *minor premiss,* and contains the Subject of the conclusion; and the other term, common to the two premises, and with which both the terms of the conclusion were separately compared, *before* they were compared with each other, is called the *middle term.* In the syllogism above, "detestable" (in the conclusion) is the major term, and "Cæsar" the minor term: hence,

Example, pointing out Major premiss, Minor premiss, and Middle Term.

"All tyrants are detestable,"

is the major premiss, and

"Cæsar was a tyrant,"

the minor premiss, and "tyrant" the middle term.

Syllogism, a mere formula.

§ 41. The syllogism, therefore, is a mere formula for ascertaining what may, or what may not, be predicated of a subject. It accomplishes this end by means of two propositions, viz. by comparing the given predicate of the first (a Major Premiss), and the given subject of the second (a Minor Premiss), respectively with one and the same third term (called the middle term), and thus—under *certain conditions*, or laws of the syllogism—to be hereafter stated—eliciting the truth (conclusion) that the given predicate must be predicated of that subject. It will be seen that the Major Premiss always declares, in a *general* way, such a relation between the Major Term and the Middle Term; and the Minor Premiss declares, in a more *particular* way, such a relation between the Minor Term and the Middle Term, as that, in the Conclusion, the Minor Term must be put under the Major Term; or in other words, that the Major Term must be predicated of the Minor Term.

How applied.

Use of the Major premiss.

Of the Minor.

Of the Middle Term.

## ANALYTICAL OUTLINE OF DEDUCTION.

Reasoning defined.

§ 42. In every instance in which we *reason*, in the strict sense of the word, that is, make use of arguments, whether for the sake of refuting an adversary, or of conveying instruction, or of satisfying our own minds on any point, whatever may be the subject we are engaged on, a certain process takes place in the mind, which is one and the same in all cases (provided it be correctly conducted), whether we use the inductive process or the deductive formulas.

The process, in all cases, the same.

Every one not conscious of the process.

Of course it cannot be supposed that every one is even conscious of this process in his own mind; much less, is competent to explain the principles on which it proceeds. This indeed is, and cannot but be, the case with every other process respecting which any system has been formed; the practice not only may exist independently of the theory, but *must* have preceded the theory. There must have been Language before a system of Grammar could be devised; and musical compositions, previous to the science of Music. This, by the way, serves to expose the futility of the popular objection against Logic; viz. that men may reason very well who know nothing of it. The parallel instances adduced show that such an objection may be urged

The same for every other process.

Elements and knowledge of elements, must precede generalization and classification of principles.

in many other cases, where its absurdity would be obvious; and that there is no ground for deciding thence, either that the system has no tendency to improve practice, or that even if it had not, it might not still be a dignified and interesting pursuit.

Logic of value.

Sameness of the reasoning process should be kept in mind.

§ 43. One of the chief impediments to the attainment of a just view of the nature and object of Logic, is the not fully understanding, or not sufficiently keeping in mind the SAMENESS of the reasoning process in all cases. If, as the ordinary mode of speaking would seem to indicate, mathematical reasoning, and theological, and metaphysical, and political, &c., were essentially different from each other, that is, different *kinds of reasoning*, it would follow, that supposing there could be at all any such science as we have described Logic, there must be so many different species or at least different branches of Logic. And such is perhaps the most prevailing notion. Nor is this much to be wondered at; since it is evident to all, that some men converse and write, in an argumentative way, very justly on one subject, and very erroneously on another, in which again others excel, who fail in the former.

All kinds of reasoning are alike in principle.

Reason of the prevailing errors.

This error may be at once illustrated and re-

The reason of the error illustrated by example, which shows that the reasoning process is always the same.

moved, by considering the parallel instance of Arithmetic; in which every one is aware that the process of a calculation is not affected by the nature of the objects whose numbers are before us; but that, for example, the multiplication of a number is the very same operation, whether it be a number of men, of miles, or of pounds; though, nevertheless, persons may perhaps be found who are accurate in the results of their calculations relative to natural philosophy, and incorrect in those of political economy, from their different degrees of skill in the subjects of these two sciences; not surely because there are different arts of arithmetic applicable to each of these respectively.

Some view Logic as a peculiar method of reasoning:

§ 44. Others again, who are aware that the simple system of Logic may be applied to all subjects whatever, are yet disposed to view it as a peculiar method of reasoning, and not, as it is, a method of unfolding and analyzing our reasoning: whence many have been led to talk of comparing Syllogistic reasoning with Moral reasoning; taking it for granted that it is possible to reason correctly without reasoning logically; which is, in fact, as great a blunder as if any one were to mistake *grammar* for a peculiar *language*, and to suppose it possible to speak

It is the only method of reasoning correctly:

correctly without speaking grammatically. They have, in short, considered Logic as *an* art of reasoning; whereas (so far as it is an art) it is *the* art of reasoning; the logician's object being, not to lay down principles by which one *may* reason, but by which all *mūst* reason, even though they are not distinctly aware of them:—to lay down rules, not which *may* be followed with advantage, but which cannot possibly be *departed* from in sound reasoning. These misapprehensions and objections being such as lie on the very threshold of the subject, it would have been hardly possible, without noticing them, to convey any just notion of the nature and design of the logical system.

It lays down rules, not which *may*, but which *must* be followed.

Misapprehensions and objections noticed.

§ 45. Supposing it then to have been perceived that the operation of reasoning is in all cases the same, the analysis of that operation could not fail to strike the mind as an interesting matter of inquiry. And moreover, since (apparent) arguments, which are unsound and inconclusive, are so often employed, either from error or design; and since even those who are not misled by these fallacies, are so often at a loss to detect and expose them in a manner satisfactory to others, or even to themselves; it could not but appear desirable to lay down some gen-

Operation of reasoning should be analyzed:

Because such analysis is necessary to furnish the

rules for the detection of error and the discovery of truth.

eral rules of reasoning, applicable to all cases; by which a person might be enabled the more readily and clearly to state the grounds of his own conviction, or of his objection to the arguments of an opponent; instead of arguing at random, without any fixed and acknowledged principles to guide his procedure. Such rules would be analogous to those of Arithmetic, which obviate the tediousness and uncertainty of calculations in the head; wherein, after much labor, different persons might arrive at different results, without any of them being able distinctly to point out the error of the rest. A system of such rules, it is obvious, must, instead of deserving to be called the art of wrangling, be more justly characterized as the "art of cutting short wrangling," by bringing the parties to issue at once, if not to agreement; and thus saving a waste of ingenuity.

Such rules are analogous to the rules of Arithmetic.

They bring the parties, in argument, to an issue.

Every conclusion is deduced from two propositions, called Premises.

If one premiss is suppressed, it is nevertheless understood.

§ 46. In pursuing the supposed investigation, it will be found that in all deductive processes every conclusion is deduced, in reality, from two other propositions (thence called *Premises*); for though one of these may be, and commonly is, suppressed, it must nevertheless be understood as admitted; as may easily be made evident by supposing the denial of the suppressed premiss, which will at once invalidate the argument. For example; in the following syllogism:

"Whatever exhibits marks of design had an intelligent author:
The world exhibits marks of design;
*Therefore*, the world had an intelligent author:"

and is necessary to the argument, though one may not be aware of it.

if any one from perceiving that "the world exhibits marks of design," infers that "it must have had an intelligent author," though he may not be aware in his own mind of the existence of any other premiss, he will readily understand, if it be *denied* that "whatever exhibits marks of design must have had an intelligent author," that the affirmative of that proposition is necessary to the validity of the argument.

Enthymeme: a syllogism with one premiss suppressed.

§ 47. When one of the premises is suppressed (which for brevity's sake it usually is), the argument is called an Enthymeme. For example:

"The world exhibits marks of design,
*Therefore* the world had an intelligent author,"

Objections made to the *assertion* or to the *force* of the argument.

is an Enthymeme. And it may be worth while to remark, that, when the argument is in this state, the objections of an opponent are (or rather appear to be) of two kinds, viz. either objections to the *assertion* itself, or objections to its *force* as an argument. For example: in the above instance, an atheist may be conceived either de-

Example.

Both premises must be true, if the argument is sound: and when both are true, the conclusion follows.

nying that the world *does* exhibit marks of design, or denying that it *follows* from thence that it had an intelligent author. Now it is important to keep in mind that the only difference in the two cases is, that in the one the *expressed* premiss is denied, in the other the *suppressed;* for the *force as an argument* of either premiss depends on the other premiss: if both be admitted, the conclusion legitimately connected with them cannot be denied.

Premiss placed *after* the conclusion is called the *Reason*.

§ 48. It is evidently immaterial to the argument whether the conclusion be placed first or last; but it may be proper to remark, that a premiss placed *after* its conclusion is called the *Reason* of it, and is introduced by one of those conjunctions which are called causal, viz. "since," "because," &c., which may indeed be employed to designate a premiss, whether it come first or last. The illative conjunctions "therefore," &c., designate the conclusion.

Illative conjunction.

Causes of error and perplexity.

Different significations of the conjunctions.

It is a circumstance which often occasions error and perplexity, that both these classes of conjunctions have also another signification, being employed to denote, respectively, *Cause* and *Effect,* as well as *Premiss* and *Conclusion.* For example: if I say, "this ground is rich, *because* the trees on it are flourishing;" or, "the trees are flourishing, and *therefore* the soil must be rich;" I employ these conjunctions to denote the connection of *Premiss* and *Conclusion;* for it is plain that the luxuriance of the trees is not the cause of the soil's fertility, but only the cause of *my knowing* it. If again I say, "the trees flourish, *because* the ground is rich;" or "the ground is rich, and *therefore* the trees flourish," I am using the very same conjunctions to denote the connection of *cause* and *effect;* for in this case, the luxuriance of the trees being evident to the eye, would hardly need to be *proved,* but might need to be accounted for. There are, however, many cases, in which the cause is employed to *prove* the existence of its effect; especially in arguments relating to *future* events; as, for example, when from favorable weather any one argues that the crops are likely to be abundant, the *cause* and the reason, in that case, coincide; and this contributes to their being so often confounded together in other cases.

Examples where the conjunctions are used logically.

Examples where they denote cause and effect.

Many cases in which the cause and the reason are the same.

§ 49. In an argument, such as the example above given, it is, as has been said, impossible for any one, who admits both premises, to avoid admitting the conclusion. But there will be frequently an apparent connection of premises with a conclusion which does not in reality follow

In every correct argument, to admit the premiss is to admit the conclusion.

Apparent connection of premises and conclusion must not be relied on.

from them, though to the inattentive or unskilful the argument may appear to be valid; and there are many other cases in which a doubt may exist whether the argument be valid or not; that is, whether it be possible or not to admit the premises and yet deny the conclusion.

General rules for argumentation necessary.

§ 50. It is of the highest importance, therefore, to lay down some regular form to which every valid argument may be reduced, and to devise a rule which shall show the validity of every argument in that form, and consequently the unsoundness of any apparent argument which cannot be reduced to it. For example; if such an argument as this be proposed:

Example of an imperfect argument.

> "Every rational agent is accountable:
> Brutes are not rational agents;
> Therefore they are not accountable;"

or again:

2d Example.

> "All wise legislators suit their laws to the genius of their nation;
> Solon did this; therefore he was a wise legislator:"

Difficulty of detecting the error.

there are some, perhaps, who would not perceive any fallacy in such arguments, especially if enveloped in a cloud of words; and still more, when the conclusion is true, or (which comes to the same point) if they are disposed to believe it; and others might perceive indeed, but might be at a loss to explain, the fallacy. Now these (apparent) arguments exactly correspond, respectively, with the following, the absurdity of the conclusions from which is manifest:

To what these apparent arguments correspond.

"Every horse is an animal:
Sheep are not horses;
Therefore, they are not animals."

A similar example.

And:

"All vegetables grow;
An animal grows;
Therefore, it is a vegetable."

2d similar example.

These last examples, I have said, correspond exactly (considered as arguments) with the former; the question respecting the validity of an argument being, not whether the conclusion be *true*, but whether it *follows* from the premises adduced. This mode of exposing a fallacy, by bringing forward a similar one whose conclusion is obviously absurd, is often, and very advantageously, resorted to in addressing those who are ignorant of Logical rules; but to lay down such rules, and employ *them* as a test, is evidently a safer and more compendious, as well as a more philosophical mode of proceeding. To attain these, it would plainly be necessary to analyze some clear and valid arguments, and to observe in what their conclusiveness consists.

These last correspond with the former.

This mode of exposing fallacy sometimes resorted to.

To lay down rules is the best way.

§ 51. Let us suppose, then, such an examination to be made of the syllogism above mentioned:

Example of a perfect syllogism.

"Whatever exhibits marks of design had an intelligent author;
The world exhibits marks of design;
Therefore, the world had an intelligent author."

What is assumed in the first premiss. In the second premiss. What we may infer.

In the first of these premises we find it assumed universally of the *class* of "things which exhibit marks of design," that they had an intelligent author; and in the other premiss, "the world" is referred to that class as comprehended in it: now it is evident that whatever is said of the whole of a class, may be said of any thing comprehended in that class; so that we are thus authorized to say of the world, that "it had an intelligent author."

Syllogism with a negative conclusion.

Again, if we examine a syllogism with a negative conclusion, as, for example,

"Nothing which exhibits marks of design could have been produced by chance;
The world exhibits, &c.;
Therefore, the world could not have been produced by chance,"

The process of reasoning the same.

the process of reasoning will be found to be the same; since it is evident that whatever is *denied* universally of any class may be denied of any thing that is comprehended in that class.

§ 52. On further examination, it will be found that all valid arguments whatever, which are based on admitted premises, may be easily reduced to such a form as that of the foregoing syllogisms; and that consequently the principle on which they are constructed is that of the formula of the syllogism. So elliptical, indeed, is the ordinary mode of expression, even of those who are considered as prolix writers, that is, so much is implied and left to be understood in the course of argument, in comparison of what is actually stated (most men being impatient even, to excess, of any appearance of unnecessary and tedious formality of statement), that a single sentence will often be found, though perhaps considered as a single argument, to contain, compressed into a short compass, a chain of several distinct arguments. But if each of these be fully developed, and the whole of what the author intended to imply be stated expressly, it will be found that all the steps, even of the longest and most complex train of reasoning, may be reduced into the above form.

All valid arguments reducible to the syllogistic form.

Ordinary mode of expressing arguments elliptical.

But when fully developed, they may all be reduced into the above form.

§ 53. It is a mistake to imagine that Aristotle and other logicians meant to propose that this prolix form of unfolding arguments should universally supersede, in argumentative discourses,

Aristotle did not mean that every argument should be

thrown into the form of a syllogism.

the common forms of expression; and that "to reason logically," means, to state all arguments at full length in the syllogistic form; and Aristotle has even been charged with inconsistency for not doing so. It has been said that he "argues like a rational creature, and never attempts to bring his own system into practice." As well might a chemist be charged with inconsistency for making use of any of the compound substances that are commonly employed, without previously analyzing and resolving them into their simple elements; as well might it be imagined that, to speak grammatically, means, to parse every sentence we utter. The chemist (to pursue the illustration) keeps by him his tests and his method of analysis, to be employed when any substance is offered to his notice, the composition of which has not been ascertained, or in which adulteration is suspected. Now a fallacy may aptly be compared to some adulterated compound; "it consists of an ingenious mixture of truth and falsehood, so entangled, so intimately blended, that the falsehood is (in the chemical phrase) *held in solution:* one drop of sound logic is that test which immediately disunites them, makes the foreign substance visible, and precipitates it to the bottom."

That form is merely a *test* of truth.

Analogy to the chemist.

The analogy continued.

To what a fallacy may be compared.

How detected.

### ARISTOTLE'S DICTUM.

§ 54. But to resume the investigation of the principles of reasoning: the maxim resulting from the examination of a syllogism in the foregoing form, and of the application of which, every valid deduction is in reality an instance, is this:

Form of every real argument.

"That whatever is predicated (that is, affirmed or denied) *universally*, of any class of things, may be predicated, in like manner (viz. affirmed or denied), of any thing comprehended in that class."

Aristotle's dictum.

This is the principle commonly called the *dictum de omni et nullo*, for the indication of which we are indebted to Aristotle, and which is the keystone of his whole logical system. It is remarkable that some, otherwise judicious writers, should have been so carried away by their zeal against that philosopher, as to speak with scorn and ridicule of this principle, on account of its obviousness and simplicity; though they would probably perceive at once in any other case, that it is the greatest triumph of philosophy to refer many, and seemingly very various phenomena to one, or a very few, simple principles; and that the more simple and evident such a principle is, provided it be truly applicable to all the cases in question, the

What the principle is called.

What writers have said of this principle; and why.

Simplicity a test of science.

No solid objection to the principle ever urged.

greater is its value and scientific beauty. If, indeed, any principle be regarded as *not* thus applicable, *that* is an objection to it of a different kind. Such an objection against Aristotle's dictum, no one has ever attempted to *establish* by any kind of proof; but it has often been *taken for granted;* it being (as has been stated) very commonly supposed, without examination, that the syllogism is a *distinct kind of argument,* and that the rules of it accordingly do not apply, nor were intended to apply, to *all* reasoning whatever, where the premises are granted or known.

What has been taken for granted.

Syllogism not a distinct kind of argument; but a form applicable to all cases.

Objection: that the syllogism was intended to make a demonstration plainer:

§ 55. One objection against the dictum of Aristotle it may be worth while to notice briefly, for the sake of setting in a clearer light the real character and object of that principle. The application of the principle being, as has been seen, to a regular and conclusive syllogism, it has been urged that the dictum was intended to prove and *make evident* the conclusiveness of such a syllogism; and that it is unphilosophical to attempt giving a *demonstration of a demonstration.* And certainly the charge would be just, if we could imagine the logician's object to be, to *increase the certainty* of a conclusion, which we are supposed to have already arrived at by the clearest possible mode

to increase the certainty of a conclusion.

of proof. But it is very strange that such an idea should ever have occurred to one who had even the slightest tincture of natural philosophy; for it might as well be imagined that a natural philosopher's or a chemist's design is to strengthen the testimony of our senses by *à priori* reasoning, and to convince us that a stone when thrown will fall to the ground, and that gunpowder will explode when fired; because they show according to their principles those phenomena must take place as they do. But it would be reckoned a mark of the grossest ignorance and stupidity not to be aware that their object is not to *prove* the existence of an individual phenomenon, which our eyes have witnessed, but (as the phrase is) to *account* for it; that is, to show according to what *principle* it takes place; to refer, in short, the *individual case* to a *general law* of nature. The object of Aristotle's dictum is precisely analogous: he had, doubtless, no thought of adding to the force of any individual syllogism; his design was to point out the *general principle* on which that process is conducted which takes place in each syllogism. And as the Laws of nature (as they are called) are in reality merely *generalized facts*, of which all the phenomena coming under them are particular instances; so, the proof drawn from

This view is entirely erroneous.

Illustration.

The object is not to prove, but to account for.

The object of the Dictum to point out the *general process* to which each case conforms.

Laws of nature, generalized facts.

The Dictum a condensed form of all demonstration.

Aristotle's dictum is not a distinct demonstration brought to confirm another demonstration, but is merely a generalized and abstract statement of *all* demonstration whatever; and is, therefore, in fact, *the very demonstration* which, under proper suppositions, accommodates itself to the various subject-matters, and which is actually employed in each particular case.

How to trace the abstracting and reasoning process.

§ 56. In order to trace more distinctly the different steps of the abstracting process, by which any particular argument may be brought into the most general form, we may first take a syllogism, that is, an argument stated accurately and at full length, such as the example formerly given:

An argument stated at full length.

"Whatever exhibits marks of design had an intelligent author;
The world exhibits marks of design;
Therefore, the world had an intelligent author:"

Propositions expressed by abstract terms.

and then somewhat generalize the expression, by substituting (as in Algebra) arbitrary unmeaning symbols for the significant terms that were originally used. The syllogism will then stand thus:

"Every B is A; C is B; therefore C is A."

The reasoning no less valid,

The reasoning, when thus stated, is no less evidently valid, whatever terms A, B, and C respectively may be supposed to stand for; such terms may indeed be inserted as to make all or some of the assertions *false;* but it will still be no less impossible for any one who *admits* the truth of the *premises,* in an argument thus constructed, to deny the conclusion; and this it is that constitutes the conclusiveness of an argument.

and equally general.

Viewing, then, the syllogism thus expressed, it appears clearly that "A stands for *any thing whatever* that is affirmed of a certain entire class" (viz. of *every* B), "which class comprehends or contains in it *something else,*" viz. C (of which B is, in the second premiss, affirmed); and that, consequently, the first term (A) is, in the conclusion, predicated of the third (C).

Syllogism so viewed, affirms general relations between the terms.

§ 57. Now, to assert the validity of this process now before us, is to state the very dictum we are treating of, with hardly even a verbal alteration, viz.:

Another form of stating the dictum.

1. Any thing whatever, predicated of a whole class;

The three things implied.

2. Under which class something else is contained;

3. May be predicated of that which is so contained.

The three members into which the maxim is here distributed, correspond to the three propo-

These three members correspond to the three propositions

sitions of the syllogism to which they are intended respectively to apply.

Advantage of substituting arbitrary symbols for the terms.

The advantage of substituting for the terms, in a regular syllogism, arbitrary, unmeaning symbols, such as letters of the alphabet, is much the same as in geometry: the reasoning itself is then considered, by itself, clearly, and without any risk of our being misled by the truth or falsity of the conclusion; which is, in fact, accidental and variable; the essential point being, as far as the *argument* is concerned, the *connection between* the premises and the conclusion. We are thus enabled to embrace the general principle of deductive reasoning, and to perceive its applicability to an indefinite number of individual cases. That Aristotle, therefore, should have been accused of making use of these symbols for the purpose of *darkening* his demonstrations, and that too by persons not unacquainted with geometry and algebra, is truly astonishing.

Connection, the essential point of the argument.

Aristotle right in using these symbols.

Syllogism equally true when abstract terms are used.

§ 58. It belongs, then, exclusively to a syllogism, properly so called (that is, a valid argument, so stated that its conclusiveness is evident from the mere *form* of the expression), that if letters, or any other unmeaning symbols, be substituted for the several terms, the validity of the argument shall still be evident. Whenever this is not the case, the supposed argument is either unsound and sophistical, or else may be reduced (without any alteration of its meaning) into the syllogistic form; in which form, the test just mentioned may be applied to it.

When not so, the supposed argument is unsound.

§ 59. What is called an unsound or fallacious argument, that is, an *apparent* argument, which is, in reality, none, cannot, of course, be reduced into this form; but when stated in the form most nearly approaching to this that is possible, its fallaciousness becomes more evident, from its nonconformity to the foregoing rule. For example:

Definition of an unsound argument.

When reduced to the form, the fallacy is more evident.

"Whoever is capable of deliberate crime is responsible;
An infant is not capable of deliberate crime;
Therefore, an infant is not responsible."

Example.

Here the term "responsible" is affirmed universally of "those capable of deliberate crime;" it might, therefore, according to Aristotle's dictum, have been affirmed of any thing contained under that class; but, in the instance before us, nothing is mentioned as contained under that class; only, the term "infant" is *excluded* from that class; and though what is affirmed of a whole class may be affirmed of any thing that *is* contained under it, there is no ground for supposing that it may be *denied* of whatever is *not*

Analysis of this syllogism.

Its defective nature pointed out.

Why the argument is not good.

so contained; for it is evidently possible that it may be applicable to a whole class and to something else besides. To say, for example, that all trees are vegetables, does not imply that *nothing else* is a vegetable. Nor, when it is said, that all who are capable of deliberate crime are responsible, does this imply that no others are responsible; for though this may be very *true*, it has not been asserted in the premiss before us; and in the analysis of an argument, we are to discard all consideration of what *might* be asserted; contemplating only what *actually is* laid down in the premises. It is evident, therefore, that such an apparent argument as the above does not comply with the rule laid down, nor can be so stated as to comply with it, and is consequently invalid.

What the statement implies.

What is to be done in the analysis of an argument.

The one above did not comply with the rule.

§ 60. Again, in this instance:

Another example.

"Food is necessary to life;
Corn is food;
Therefore corn is necessary to life:"

In what the argument is defective.

the term "necessary to life" is affirmed of food, but *not universally;* for it is not said of *every kind of food:* the meaning of the assertion being manifestly that *some* food is necessary to life: here again, therefore, the rule has not been complied with, since that which has been predicated (that is, affirmed or denied), not of the *whole*, but of a *part* only of a certain class, cannot be, on that ground, predicated of whatever is contained under that class.

Why we cannot predicate of corn what was predicated of food.

DISTRIBUTION AND NON-DISTRIBUTION OF TERMS.

Fallacy in the last example.

§ 61. The fallacy in this last case is, what is usually described in logical language as consisting in the "non-distribution of the middle term;" that is, its not being employed to denote *all* the objects to which it is applicable. In order to understand this phrase, it is necessary to observe, that a term is said to be "distributed," when it is taken universally, that is, so as to stand for all its significates; and consequently "undistributed," when it stands for only a portion of its significates.* Thus, "*all* food," or *every* kind of food, are expressions which imply the distribution of the term "food;" "*some* food" would imply its non-distribution.

Non-distribution of the middle term.

What *distribution* means.

Non-distribution.

Now, it is plain, that if in each premiss a *part* only of the middle term is employed, that is, if it be not at all distributed, no conclusion can be drawn. Hence, if in the example formerly adduced, it had been merely stated that "*some-*

How the example might have been varied.

* Section 15.

What it would then have implied.

*thing*" (not "*whatever*," or "*every thing*") "which exhibits marks of design, is the work of an intelligent author," it would not have followed, from the world's exhibiting marks of design, that that is the work of an intelligent author.

Words marking distribution or non-distribution not always expressed.

§ 62. It is to be observed also, that the words "all" and "every," which mark the distribution of a term, and "some," which marks its non-distribution, are not always expressed: they are frequently understood, and left to be supplied by the context; as, for example, "food is necessary;" viz. "*some* food;" "man is mortal;" viz. "*every* man." Propositions thus expressed are called by logicians "*indefinite*," because it is left undetermined by the form of the expression whether the subject be distributed or not. Nevertheless it is plain that in every proposition the subject either is or is not meant to be distributed, though it be not declared whether it is or not; consequently, every proposition, whether expressed indefinitely or not, must be understood as either "universal" or "particular;" those being called universal, in which the predicate is said of the whole of the subject (or, in other words, where all the significates are included); and those particular, in which only a part of them is included. For example:

Such propositions are called Indefinite.

But every proposition must be either Universal or Particular.

Example of each.

"All men are sinful," is universal: "some men are sinful," particular; and this division of propositions, having reference to the distribution of the *subject*, is, in logical language, said to be according to their "*quantity*."

This division relates to quantity.

§ 63. But the distribution or non-distribution of the *predicate* is entirely independent of the *quantity* of the proposition; nor are the signs "all" and "some" ever affixed to the predicate; because its distribution depends upon, and is indicated by, the "*quality*" of the proposition; that is, its being *affirmative* or *negative*; it being a universal rule, that the predicate of a negative proposition is distributed, and of an affirmative, undistributed. The reason of this may easily be understood, by considering that a term which stands for a whole class may be applied to (that is, *affirmed* of) any thing that is comprehended under that class, though the term of which it is thus affirmed may be of much narrower extent than that other, and may therefore be far from coinciding with the *whole* of it. Thus it may be said with truth, that "the Negroes are uncivilized," though the term "uncivilized" be of much wider extent than "Negroes," comprehending, besides them, Patagonians, Esquimaux, &c.; so that it would not be allowable to assert, that

Distribution of the predicate has no reference to *quantity*.

Has reference to *quality*.

When it is distributed:

The reason of this.

The predicate of affirmative propositions may be applicable to the subject, and yet of much wider extent.

Hence, only a part of the term is used.

*all* who are uncivilized are Negroes." It is evident, therefore, that it is a *part* only of the term "uncivilized" that has been affirmed of "Negroes;" and the same reasoning applies to every affirmative proposition.

But it may be of equal extent with the subject:

this not implied in the form of the expression.

It may indeed so happen, that the subject and predicate coincide, that is, are of equal extent; as, for example: "all men are rational animals;" "all equilateral triangles are equiangular;" (it being equally true, that "all rational animals are men," and that "all equiangular triangles are equilateral;") yet this is not *implied by the form of the expression;* since it would be no less true that "all men are rational animals," even if there were other rational animals besides men.

If any part of the predicate is applicable to the subject, it may be affirmed of the subject.

If a predicate is *denied* of a subject, the whole predicate is denied of the subject.

It is plain, therefore, that if *any part* of the predicate is applicable to the subject, it may be affirmed, and of course cannot be denied, of that subject; and consequently, when the predicate *is denied* of the subject, this implies that *no part* of that predicate is applicable to that subject; that is, that the *whole* of the predicate is denied of the subject: for to say, for example, that "no beasts of prey ruminate," implies that beasts of prey are excluded from the *whole class* of ruminant animals, and consequently that "no ruminant animals are beasts of prey." And hence results the above-mentioned rule, that the distribution of the predicate is implied in negative propositions, and its non-distribution in affirmatives.

Distribution of predicate implied in negative propositions: non-distribution in affirmatives.

§ **64.** It is to be remembered, therefore, that it is not sufficient for the middle term to *occur* in a universal proposition; since if that proposition be an affirmative, and the middle term be the *predicate* of it, it will not be distributed. For example: if in the example formerly given, it had been merely asserted, that "all the works of an intelligent author show marks of design," and that "the universe shows marks of design," nothing could have been proved; since, though both these propositions are universal, the middle term is made the predicate in each, and both are affirmative; and accordingly, the rule of Aristotle is not here complied with, since the term "work of an intelligent author," which is to be proved applicable to "the universe," would not have been affirmed of the middle term ("what shows marks of design") under which "universe" is contained; but the middle term, on the contrary, would have been affirmed of it.

Not sufficient for the middle term to occur in a universal proposition.

It must be so connected with the terms of the conclusion, that those terms may be compared together.

If, however, one of the premises be negative, the middle term may then be made the predicate

If one premiss be negative, the middle term may be made the predicate of that, and will be distributed.

of that, and will thus, according to the above remark, be distributed. For example:

> "No ruminant animals are predacious:
> The lion is predacious;
> Therefore the lion is not ruminant:"

this is a valid syllogism; and the middle term (predacious) is distributed by being made the predicate of a negative proposition. The form, indeed, of the syllogism is not that prescribed by the dictum of Aristotle, but it may easily be reduced to that form, by stating the first proposition thus: "No predacious animals are ruminant;" which is manifestly implied (as was above remarked) in the assertion that "no ruminant animals are predacious." The syllogism will thus appear in the form to which the dictum applies.

The form of this syllogism will not be that prescribed by the dictum, but may be reduced to it.

All arguments cannot be reduced by so short a process.

§ 65. It is not every argument, indeed, that can be reduced to this form by so short and simple an alteration as in the case before us. A longer and more complex process will often be required, and rules may be laid down to facilitate this process in certain cases; but there is no sound argument but what *can* be reduced into this form, without at all departing from the real meaning and drift of it; and the form will be

But all arguments may

found (though more prolix than is needed for ordinary use) the most perspicuous in which an argument can be exhibited.

be reduced to the prescribed form.

§ 66. All deductive reasoning whatever, then, rests on the one simple principle laid down by Aristotle, that

All deductive reasoning rests on the dictum.

"What is predicated, either affirmatively or negatively, of a term distributed, may be predicated in like manner (that is, affirmatively or negatively) of any thing contained under that term."

So that, when our object is to prove any proposition, that is, to show that one term may rightly be affirmed or denied of another, the process which really takes place in our minds is, that we *refer* that term (of which the other is to be thus predicated) to some class (that is, middle term) of which that other may be affirmed, or denied, as the case may be. Whatever the subject-matter of an argument may be, the reasoning itself, considered by itself, is in every case the same process; and if the writers against Logic had kept this in mind, they would have been cautious of expressing their contempt of what they call "syllogistic reasoning," which embraces *all* deductive reasoning; and instead of ridiculing Aristotle's principle for its obviousness and simplicity, would have perceived that these are, in fact, its

What are the processes of proof.

The reasoning always the same.

Mistakes of writers on Logic.

Aristotle's principle

simple and general. highest praise: the easiest, shortest, and most evident theory, provided it answer the purpose of explanation, being ever the best.

### RULES FOR EXAMINING SYLLOGISMS.

Tests of the validity of syllogisms. § 67. The following axioms or canons serve as tests of the validity of that class of syllogisms which we have considered.

1st test. 1st. *If two terms agree with one and the same third, they agree with each other.*

2d test. 2d. *If one term agrees and another disagrees with one and the same third, these two disagree with each other.*

The first the test of all affirmative conclusions. The second of negative. On the former of these canons rests the validity of *affirmative* conclusions; on the latter, of *negative:* for, no syllogism can be faulty which does not violate these canons; none correct which does; hence, on these two canons are built the following rules or cautions, which are to be observed with respect to syllogisms, for the purpose of ascertaining whether those canons have been strictly observed or not.

Every syllogism has three and only three terms. 1st. *Every syllogism has three and only three terms;* viz. the middle term and the two terms of the Conclusion: the terms of the Conclusion are sometimes called extremes.

Every syllo- 2d. *Every syllogism has three and only three*

*propositions;* viz. the major premiss; the minor premiss; and the conclusion.

gism has three and only three propositions.

3d. *If the middle term is ambiguous, there are in reality two middle terms, in sense,* though but one in *sound.*

Middle term must not be ambiguous.

There are two cases of ambiguity: 1st. Where the middle term is *equivocal;* that is, when used in different senses in the two premises. For example:

Two cases.

1st case.

> "*Light* is contrary to darkness;
> Feathers are *light;* therefore,
> Feathers are contrary to darkness."

Example.

2d. Where the middle term is *not* distributed; for as it is then used to stand for a *part* only of its *significates,* it may happen that one of the extremes is compared with *one part* of the whole term, and the other with another part of it. For example:

2d case.

> "White is a color;
> Black is a color; therefore,
> Black is white."

Again:

> "Some animals are beasts;
> Some animals are birds; therefore,
> Some birds are beasts."

Examples.

3d. *The middle term, therefore, must be distributed, once, at least,* in the premises; that is,

The middle term must be once distributed;

and once is sufficient.

by being the subject of a universal,* or predicate of a negative;† and once is sufficient; since if one extreme has been compared with a *part* of the middle term, and another to the *whole* of it, they must have been compared with the same.

No term must be distributed in the conclusion which was not distributed in a premiss.

4th. *No term must be distributed in the conclusion which was not distributed in one of the premises;* for, that would be to employ the *whole* of a term in the conclusion, when *you* had employed only a *part* of it in the premiss; thus, in reality, to introduce a fourth term. This is called an *illicit* process either of the major or minor term.‡ For example:

Example.

> "All quadrupeds are animals,
> A bird is not a quadruped; therefore,
> It is not an animal." Illicit process of the major.

Negative premises prove nothing.

5th. *From negative premises you can infer nothing.* For, in them the Middle is pronounced *to disagree with both extremes;* therefore they cannot be compared together: for, the extremes can only be compared when the middle agrees with *both;* or, agrees with *one*, and disagrees with the *other*. For example:

Example.

> "A fish is not a quadruped;"
> "A bird is not a quadruped," proves nothing.

---

* Section 62. † Section 63. ‡ Section 40.

6th. *If one premiss be negative, the conclusion must be negative;* for, in that premiss the middle term is pronounced to disagree with one of the extremes, and in the other premiss (which of course is affirmative by the preceding rule), to agree with the other extreme; therefore, the extremes disagreeing with each other, the conclusion is negative. In the same manner it may be shown, that *to prove a negative conclusion, one of the premises must be a negative.*

If one premiss is negative, the conclusion will be negative;

and reciprocally.

By these six rules all Syllogisms are to be tried; and from them it will be evident, 1st, that *nothing can be proved from two particular premises;* (since you will then have *either the middle term undistributed,* or an *illicit process.* For example:

What follows from these six rules.

> "Some animals are sagacious;
> Some beasts are not sagacious;
> Some beasts are not animals.")

And, for the same reason, 2dly, that if one of the premises be particular, the conclusion must be particular. For example:

2d inference.

> "All who fight bravely deserve reward;
> "Some soldiers fight bravely;" you can only infer that
> "Some soldiers deserve reward:"

Example.

for to infer a universal conclusion would be an illicit process of the minor. But from two

Two universal premises do not always give a universal conclusion.

universal Premises you cannot always infer a universal Conclusion. For example:

> "All gold is precious;
> All gold is a mineral; therefore,
> *Some* mineral is precious.'

And even when we *can* infer a universal, we are always *at liberty* to infer a particular; since what is predicated of all may *of course* be predicated of some.

### OF FALLACIES.

Definition of a fallacy.

§ 68. By a fallacy is commonly understood "any unsound mode of arguing, which appears to demand our conviction, and to be decisive of the question in hand, when in fairness it is not." In the practical detection of each individual fallacy, much must depend on natural and acquired acuteness; nor can any rules be given, the mere learning of which will enable us to *apply* them with mechanical certainty and readiness; but still we may give some hints that will lead to correct general views of the subject, and tend to *engender such a habit of mind*, as will lead to critical examinations.

Detection of, depends on acuteness.

Hints and rules useful.

Same of Logic in general.

Indeed, the case is the same with respect to Logic in general; scarcely any one would, in ordinary practice, state to himself either his own or another's reasoning, in syllogisms at full length; yet a familiarity with logical principles tends very much (as all feel, who are really well acquainted with them) to beget a habit of clear and sound reasoning. The truth is, in this as in many other things, there are processes going on in the mind (when we are practising any thing quite familiar to us), with such rapidity as to leave no trace in the memory; and we often apply principles which did not, as far as we are conscious, even occur to us at the time.

Logic tends to cultivate habits of clear reasoning.

The habit fixed, we naturally follow the processes.

§ 69. Let it be remembered, that in every process of reasoning, logically stated, the conclusion is inferred from two antecedent propositions, called the Premises. Hence, it is manifest, that in every argument, the fault, if there be any, must be either,

Conclusion follows from two antecedent premises.

Fallacy, if any, either in the premises

1st. In the premises; or,

2d. In the conclusion (when it does not follow from them); or,

or conclusion, or both.

3d. In both.

In every fallacy, the conclusion either *does* or *does not follow from the premises.*

When the fault is in the premises; that is, when they are such as ought not to have been assumed, and the conclusion legitimately follows from them, the fallacy is called a Material Fal-

When in the premises:

lacy, because it lies in the matter of the argument.

When in the conclusion. Where the conclusion does not follow from the premises, it is manifest that the fault is in the *reasoning*, and in that alone: these, therefore, are called Logical Fallacies, as being properly violations of those rules of reasoning which it is the province of logic to lay down.

When in both. When the fault lies in both the premises and reasoning, the fallacy is both Material and Logical.

Rules for examining a train of argument. § 70. In examining a train of argumentation, to ascertain if a fallacy have crept into it, the following points would naturally suggest themselves:

1st Rule. 1st. What is the proposition to be proved? On what facts or truths, as premises, is the argument to rest? and, What are the marks of truth by which the conclusion may be known?

2d Rule. 2d. Are the premises both true? If facts, are they substantiated by sufficient proofs? If truths, were they logically inferred, and from correct premises?

3d Rule. 3d. Is the middle term what it should be, and the conclusion logically inferred from the premises?

Suggestions serve as guides, These general suggestions may serve as guides in examining arguments for the purpose of detecting fallacies; but however perfect general rules may be, it is quite certain that error, in its thousand forms, will not always be separated from truth, even by those who most thoroughly understand and carefully apply such rules.

to detect error.

## CONCLUDING REMARKS.

§ 71. The imperfect and irregular sketch which has here been attempted of deductive logic, may suffice to point out the general drift and purpose of the science, and to show its entire correspondence with the reasonings in Geometry. The analytical form, which has here been adopted, is, generally speaking, better suited for *introducing* any science in the plainest and most interesting form; though the synthetical is the more regular, and the more compendious form for storing it up in the memory.

Logic corresponds with the reasonings in Geometry.

Analytical form.

Synthetical form.

§ 72. It has been a matter about which writers on logic have differed, whether, and in conformity to what principles, Induction forms a part of the science; Archbishop Whately maintaining that logic is only concerned in inferring truths from known and admitted premises, and that all reasoning, whether Inductive or Deductive, is shown by analysis to have the syllogism

Induction: does it form a part of Logic?

Whately's opinion:

Mill's views. for its type; while Mr. Mill, a writer of perhaps greater authority, holds that deductive logic is but the carrying out of what induction begins; that all reasoning is founded on principles of inference ulterior to the syllogism, and that the syllogism is the test of deduction only.

Reasons for the course taken. Without presuming at all to decide definitively a question which has been considered and passed upon by two of the most acute minds of the age, it may perhaps not be out of place to state the reasons which induced me to adopt the opinions of Mr. Mill in view of the particular use which I wished to make of logic.

Leading objects of the plan: § 73. It was, as stated in the general plan, one of my leading objects to point out the correspondence between the science of logic and the science of mathematics: to show, in fact, that mathematical reasoning conforms, in every respect, to the strictest rules of logic, and is indeed but logic applied to the abstract quantities, Number and Space. In treating of space, about which the science of Geometry is conversant, we shall see that the reasoning rests mainly on the axioms, and that these are established by inductive processes. The processes of reasoning which relate to numbers, whether the numbers are represented by figures or letters, consist of two parts.

To show that mathematical reasoning conforms to logical rules.

Axioms, how established.

1st. To obtain formulas for, that is, to express in the language of science, the relations between the quantities, facts, truths or principles, whatever they may be, that form the subject of the reasoning; and,

Two parts of the reasoning process.

2dly. To deduce from these, by processes purely logical, all the truths which are implied in them, as premises.

§ 74. Before dismissing the subject, it may be well to remark, that every induction may be thrown into the form of a syllogism, by supplying the major premiss. If this be done, we shall see that something equivalent to *the uniformity of the course of nature* will appear as the ultimate major premiss of all inductions; and will, therefore, stand to all inductions in the relation in which, as has been shown, the major premiss of a syllogism always stands to the conclusion; not contributing at all to prove it, but being a necessary condition of its being proved. This fact sustains the view taken by Mr. Mill, as stated above; for, this ultimate major premiss, or any substitution for it, is an inference by Induction, but cannot be arrived at by means of a syllogism.

All Induction may be thrown into the form of the Syllogism, by admitting a proper major premiss.

How this major premiss is obtained.

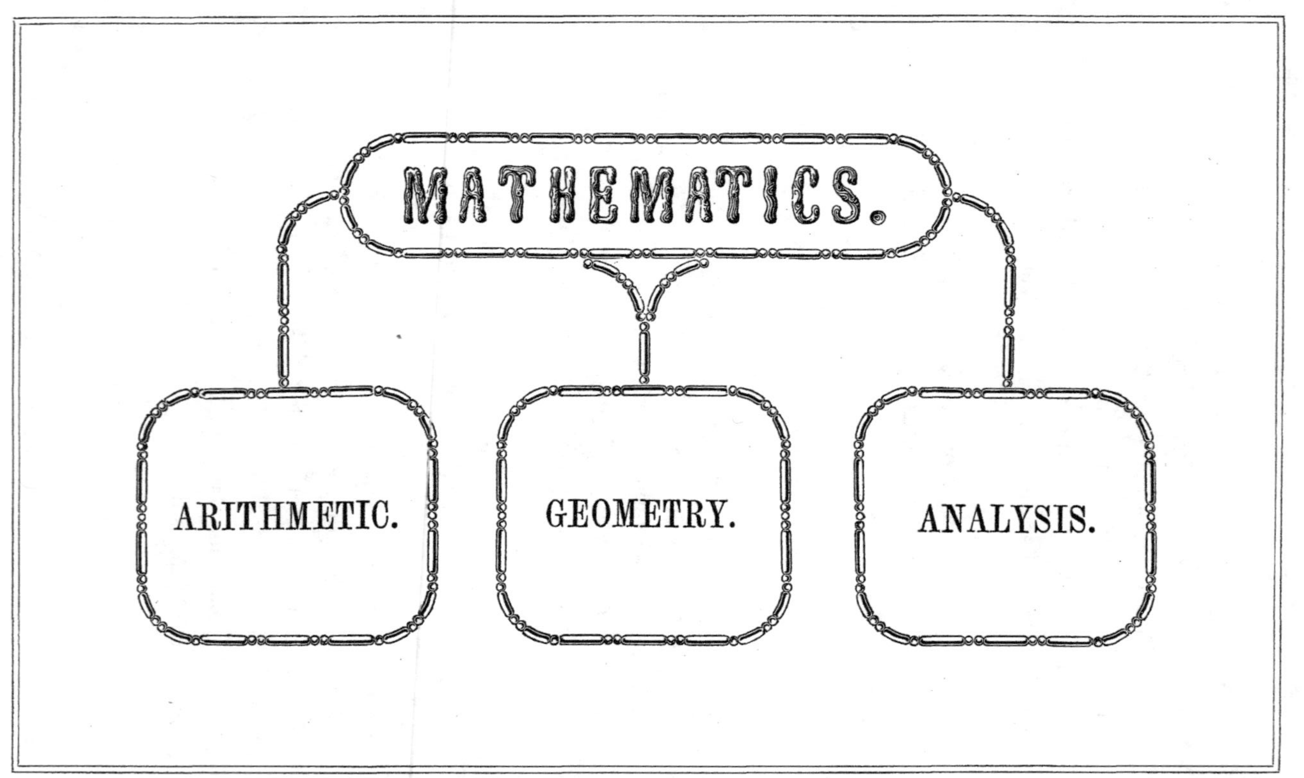
MATHEMATICS.
ARITHMETIC.
GEOMETRY.
ANALYSIS.

# BOOK II.

## MATHEMATICAL SCIENCE.

### CHAPTER I.

QUANTITY AND MATHEMATICAL SCIENCE DEFINED—DIFFERENT KINDS OF QUANTITY—LANGUAGE OF MATHEMATICS EXPLAINED—SUBJECTS CLASSIFIED—UNIT OF MEASURE DEFINED—MATHEMATICS A DEDUCTIVE SCIENCE.

#### QUANTITY.

§ 75. QUANTITY is a general term applicable to every thing which can be increased or diminished, and measured. There are two kinds of quantity: *Quantity defined.*

1st. Abstract Quantity, or quantity, the conception of which does not involve the idea of matter; and, *Abstract.*

2dly. Concrete Quantity, which embraces every thing that is material. *Concrete.*

§ 76. Mathematics is the science of quantity; that is, the science which treats of the measures of quantities and their relations to each other. It is divided into two parts: *Mathematics defined.*

Pure Mathematics. 1st. The Pure Mathematics, embracing the principles of the science, and all explanations of the processes by which those principles are derived from the laws of the abstract quantities, Number and Space; and,

Mixed Mathematics. 2d. The Mixed Mathematics, embracing the applications of those principles to all investigations and to the solution of all questions of a practical nature, whether they relate to abstract or concrete quantity.

Mathematics, as used by the ancients: embraced all subjects which were disciplinary in their nature. § 77. Mathematics, in its primary signification, as used by the ancients, embraced every acquired science, and was equally applicable to all branches of knowledge. Subsequently it was restricted to those branches only which were acquired by severe study, or discipline, and its votaries were called Disciples. Those subjects, therefore, which required patient investigation, exact reasoning, and the aid of the mathematical analysis, were called Disciplinal or Mathematical, because of the greater evidence in the arguments, the infallible certainty of the conclusions, and the mental training and development which such exercises produced.

Pure Mathematics, § 78. It has already been observed that the pure Mathematics embrace all the principles of the science, and that these principles are deduced, by processes of reasoning upon the two abstract quantities, Number and Space. All the definitions and axioms, and all the truths deduced from them, are traceable to those two sources. Here, then, two important questions present themselves:

What they embrace: relate to Number and Space.

Two questions.

1st. How are we to attain a clear and true conception of these quantities? and,

How do we conceive of the quantities?

2dly. How are we to represent them, and what *language* are we to employ, so as to make their properties and relations subjects of investigation?

How represent them.

## NUMBER.

§ 79. NUMBERS are expressions for one or more things of the same kind. How do we attain unto the significations of such expressions? By first presenting to the mind, through the eye, a single thing, and calling it ONE. Then presenting two things, and naming them TWO: then three things, and naming them THREE; and so on for other numbers. Thus, we acquire primarily, in a *concrete* form, our elementary notions of number, by perception, comparison, and reflection; for, we must first perceive how many things are numbered; then compare what is designated by the word one, with what is

Number defined.

How we obtain an idea of number.

It is done by perception, comparison, and reflection.

Reasons.

designated by the words two, three, &c., and then reflect on the results of such comparisons until we clearly apprehend the difference in the signification of the words. Having thus acquired, in a concrete form, our conceptions of numbers, we can consider numbers as separated from any particular objects, and thus form a conception of them in the abstract. We require but two axioms for the formation of all numbers:

Two axioms necessary for the formation of numbers.

1st axiom.

1st. That one may be added to any number, and that the number which results will be greater by one than the number to which the one was added.

2d axiom.

2d. That one may be divided into any number of equal parts.

Language employed.

§ 80. But what language are we to employ as best suited to furnish instruments of thought, and the means of recording our ideas and expressing them to others? The ten characters, called figures, are the alphabet of this language, and the various ways in which they are combined will be fully explained under the head Arithmetic, a chapter devoted to the consideration of numbers, their laws and language.

The ten figures are its alphabet.

## SPACE.

§ 81. SPACE is indefinite extension. We acquire our ideas of it by observing that *parts* of it are occupied by matter or bodies. This enables us to attach a definite idea to the word *place*. We are then able to say, intelligibly, that a point is that which has place, or position in space, without occupying any part of it. Having conceived a second point in space, we can understand the important axiom, "A straight line is the shortest distance between two points;" and this line we call *length* or a *dimension* of space.

Space defined.

Place: a point.

Axiom concerning the straight line.

§ 82. If we conceive a second straight line to be drawn, meeting the first, but lying in a direction directly from it, we shall have a second dimension of space, which we call *breadth*. If these lines be prolonged, in both directions, they will include four portions of space, which make up what is called a plane surface or plane: hence, a plane has two dimensions, length and breadth. If now we draw a line on either side of this plane, we shall have another dimension of space, called *thickness:* hence, space has three dimensions—length, breadth, and thickness.

Breadth defined.

A plane defined.

Space has three dimensions.

Figure defined. § 83. A portion of space limited by boundaries, is called a *Figure.* If such portion of space Line defined have but one dimension, it is called a line, and may be limited by two points, one at each extremity. Two kinds of lines: straight and curved. There are two kinds of lines, straight and curved. A straight line, is one which does not change its direction between any two of its points, and a curved line constantly changes its direction at every point.

Surface: § 84. A portion of space having two dimensions is called a surface. Plane, Curved. There are two kinds of surfaces—Plane Surfaces and Curved Surfaces. Difference. With the former, a straight line, having two points in common, will always coincide, however it may be placed, while with the latter it will not. Boundaries of a surface. The boundaries of surfaces are lines, straight or curved.

Solid defined. § 85. A portion of space having three dimensions, is called a solid, and solids are bounded either by plane or curved surfaces.

§ 86. The definitions and axioms relating to space, and all the reasonings founded on them, Science of Geometry. make up the science of Geometry. They will all be fully treated under that head.

## ANALYSIS.

§ 87. ANALYSIS is a general term embracing all the operations which can be performed on quantities when represented by letters. In this branch of mathematics, all the quantities considered, whether abstract or concrete, are represented by letters of the alphabet, and the operations to be performed on them are indicated by a few arbitrary signs. The letters and signs are called Symbols, and by their combination we form the Algebraic Notation and Language.

Analysis. Quantities represented by letters. Symbols.

§ 88. Analysis, in its simplest form, takes the name of Algebra; Analytical Geometry, the Differential and Integral Calculus, extended to include the Theory of Variations, are its higher and most advanced branches.

Analysis, Algebra; Analytical Geometry. Calculus.

§ 89. The term Analysis has also another signification. It denotes the process of separating any complex whole into the elements of which it is composed. It is opposed to Synthesis, a term which denotes the processes of first considering the elements separately, then combining them, and ascertaining the results of the combination.

Term Analysis defined. Its nature. Synthesis defined.

Analytical method. Synthetical method.

The Analytical method is best adapted to investigation, and the presentation of subjects in their general outlines; the Synthetical method is best adapted to instruction, because it exhibits all the parts of a subject separately, and in their proper order and connection. Analysis deduces all the parts from a whole: Synthesis forms a whole from the separate parts.

Arithmetic, Algebra, Geometry, elementary branches.

§ 90. Arithmetic, Algebra, and Geometry are the elementary branches of Mathematical Science. Every truth which is established by mathematical reasoning, is developed by an arithmetical, geometrical, or analytical process, or by a combination of them. The reasoning in each branch is conducted on principles identically the same. Every sign, or symbol, or technical word, is accurately defined, so that to each there is attached a definite and precise idea. Thus, the language is made so exact and certain, as to admit of no ambiguity.

Language exact.

## LANGUAGE OF MATHEMATICS.

Language of mathematics mixed.

§ 91. The language of Mathematics is mixed. Although composed mainly of symbols, which are defined with reference to the uses which are made of them, and therefore have a precise signification; it is also composed, in part, of words transferred from our common language. The symbols, although arbitrary signs, are, nevertheless, entirely general, as signs and instruments of thought; and when the sense in which they are used is once fixed, by definition, they preserve throughout the entire analysis precisely the same signification. The meaning of the words borrowed from our common vocabulary is often modified, and sometimes entirely changed, when the words are transferred to the language of science. They are then used in a particular sense, and are said to have a *technical* signification.

Symbols general.

Words borrowed from common language, are modified and used in a technical sense.

§ 92. It is of the first importance that the elements of the language be clearly understood, —that the signification of every word or symbol be distinctly apprehended, and that the connection between the thought and the word or symbol which expresses it be so well established that the one shall immediately suggest the other. It is not possible to pursue the subtle reasonings of Mathematics, and to carry out the trains of thought to which they give rise, without entire familiarity with those means which the mind employs to aid its investigations. The child cannot read till he has learned the alphabet;

Language must be understood:

Mathematical reasonings require it.

Cannot use any language

well till we know it.

nor can the scholar feel the delicate beauties of Shakspeare, or be moved by the sublimity of Milton, before studying and learning the language in which their immortal thoughts are clothed.

Quantities are represented by symbols; and are operated on by these symbols.

§ 93. All Quantities, whether abstract or concrete, are, in mathematical science, presented to the mind by arbitrary symbols. They are viewed and operated on through these symbols which represent them; and all operations are indicated by another class of symbols called *signs*. These, combined with the symbols which represent the quantities, make up, as we have stated above, the pure mathematical language; and this, in connection with that which is borrowed from our common language, forms the language of mathematical science. This language is at once comprehensive and accurate. It is capable of stating the most general proposition, and presenting to the mind, in their proper order, every elementary principle connected with its solution. By its generality it reaches over the whole field of the pure and mixed sciences, and gathers into condensed forms all the conditions and relations necessary to the development of particular facts and universal truths; and thus, the skill of the

Signs.

What constitutes the language.

Its nature.

What it accomplishes.

analyst deduces from the same equation the velocity of an apple falling to the ground, and the verification of the law of universal gravitation.

Extent and power of Analysis.

## QUANTITY MEASURED.

§ 94. Quantity has been defined, "any thing which can be increased or diminished, and measured." The terms increased or diminished, are easily understood, implying merely the property of being made larger or smaller. The term measured is not so easily explained, because it has only a relative meaning.

Quantity.

Increased and diminished, defined.

The term "measured," applied to a quantity, implies the existence of some known quantity of the same kind, which is regarded as a standard, and with which the quantity to be measured is compared with respect to its extent or magnitude. To such standard, whatever it may be, we give the name of *unity*, or *unit of measure;* and the number of times which any quantity contains its unit of measure, is the numerical value of the quantity measured. The extent or magnitude of a quantity is, therefore, merely relative, and hence, we can form no idea of it, except by the aid of comparison. Space, for example, is entirely indefinite, and we measure parts of it by means of certain standards, called

Measured.

What it means.

Standard: is called unity.

Magnitude: merely relative.

Space: indefinite.

Measurement ascertains relation: measures; and after any measurement is completed, we have only ascertained the *relation* or *proportion* which exists between the standard we a process of comparison. adopted and the thing measured. Hence, measurement is, after all, but a mere process of comparison.

Weight and velocity: known by comparison. § 95. The abstract quantities, Weight and Velocity, are but vague and indefinite conceptions, until compared with their units of measure, and even these are arrived at only by pro- Comparison a general method. cesses of comparison. Indeed, most of our knowledge of all subjects is obtained in the same way. We compare together, very carefully, all the facts which form the basis of an induction; and we rely on the comparison of the terms in the major and minor premises for every conclusion by a deductive process.

Quantity. § 96. Quantity, as we have seen, is divided into Abstract and Concrete—the abstract quan- Abstract. tity being a mere mental conception, having for its sign a number, a letter, or a geometrical Concrete. figure. A concrete quantity is a physical object, or a collection of such objects, and may How represented. likewise be represented by numbers, letters, or by the geometrical magnitudes regarded as ma- Example of the abstract. terial. The number "three" is entirely abstract, expressing an idea having no connection with material things; while the number "three pounds of tea," or "three apples," presents to the mind an idea of physical objects. So, a portion of space, bounded by a surface, all the points of which are equally distant from a certain point within called the centre, is but a mental conception of form; but regarded as a solid mass, it gives rise to the additional idea of a material substance.

Example of the abstract.

Of the concrete.

## PURE MATHEMATICS.

Pure Mathematics:

what are its foundations.

§ 97. The Pure Mathematics are based on definitions and intuitive truths, called axioms, which are inferred from observation and experience; that is, observation and experience furnish the information necessary to such intuitive inductions.* From these definitions and axioms, as premises, all the truths of the science are established by processes of deductive reasoning; and there is not, *in the whole range of mathematical science any logical test of truth, but in a conformity of the conclusions to the definitions and axioms, or to such principles as have been established from them.* Hence, we see, that the science of Pure Mathematics, which consists merely in inferring, by fixed rules, all the

Its tests of truths:

what they are.

In what the science consists.

* Section 27.

Is purely Deductive.

truths which can be deduced from given premises, is purely a Deductive Science. The precision and accuracy of the definitions; the certainty which is felt in the truth of the axioms; the obvious and fixed relation between the *sign* and the *thing signified;* and the certain formulas to which the reasoning processes are reduced, have given to mathematics the name of "Exact Science."

Precision of its language.

Exact Science.

All reasoning based on definitions and axioms.

§ 98. We have remarked that all the reasonings of mathematical science, and all the truths which they establish, are based on the definitions and axioms, which correspond to the major premiss of the syllogism. If the resemblance which the minor premiss asserts to the middle term were obvious to the senses, as it is in the proposition, "Socrates was a man," or were at once ascertainable by direct observation, or were as evident as the intuitive truth, "A whole is equal to the sum of all its parts;" there would be no necessity for trains of reasoning, and Deductive Science would not exist. Trains of reasoning are necessary only for the sake of extending the definitions and axioms to other cases in which we not only cannot directly observe what is to be proved, but cannot directly observe even the mark which is to prove it.

Relations not obvious.

Deductive Science, necessary.

Trains of reasoning:

what they accomplish.

§ 99. Although the syllogism is the ultimate test in all deductive reasoning (and indeed in all inductive, if we admit the uniformity of the course of nature), still we do not find it convenient or necessary, in mathematics, to throw every proposition into the form of a syllogism.

Syllogism, the final test of deduction.

The definitions and axioms, and the propositions established from them, are our tests of truth; and whenever any new proposition can be brought to conform to any one of these tests, it is regarded as proved, and declared to be true.

Axioms and definitions, tests of truth:

A proposition: when proved.

§ 100. When general formulas have been framed, determining the limits within which the deductions may be drawn (that is, what shall be the tests of truth), as often as a new case can be at once seen to come within one of the formulas, the principle applies to the new case, and the business is ended. But new cases are continually arising, which do not obviously come within any formula that will settle the questions we want solved in regard to them, and it is necessary to reduce them to such formulas. This gives rise to the existence of the science of mathematics, requiring the highest scientific genius in those who contributed to its creation, and calling for a most continued and vigorous

When a principle may be regarded as proved.

Trains of reasoning: why necessary.

They give rise to the science of mathematics.

exertion of intellect, in order to appropriate it when created.

### COMPARISON OF QUANTITIES.

Mathematics concerned with Number and Space.

§ 101. We have seen that the pure mathematics are concerned with the two abstract quantities, Number and Space. We have also seen that reasoning necessarily involves comparison: hence, mathematical reasoning must consist in comparing the quantities which come from Number and Space with each other.

Reasoning involves comparison.

Two quantities can sustain but two relations.

§ 102. Any two quantities, compared with each other, must necessarily sustain one of two relations: they must be equal or unequal. What axioms or formulas have we for inferring the one or the other?

### AXIOMS OR FORMULAS FOR INFERRING EQUALITY.

Formulas for Equality.

1. Things which being applied to each other coincide, are equal to one another.
2. Things which are equal to the same thing are equal to one another.
3. A whole is equal to the sum of all its parts.
4. If equals be added to equals, the sums are equal.

5. If equals be taken from equals, the remainders are equal.

### AXIOMS OR FORMULAS FOR INFERRING INEQUALITY.

Formulas for Inequality.

1. A whole is greater than any of its parts.

2. If equals be added to unequals, the sums are unequal.

3. If equals be taken from unequals, the remainders are unequal.

Outline of Mathematics completed.

§ 103. We have thus completed a very brief and general analytical view of Mathematical Science. We have endeavored to point out the character of the definitions, and the sources as well as the nature of the elementary and intuitive propositions on which the science rests; the kind of reasoning employed in its creation, and its divisions resulting from the use of different symbols and differences of language. We shall now proceed to treat the subjects separately.

What features have been sketched.

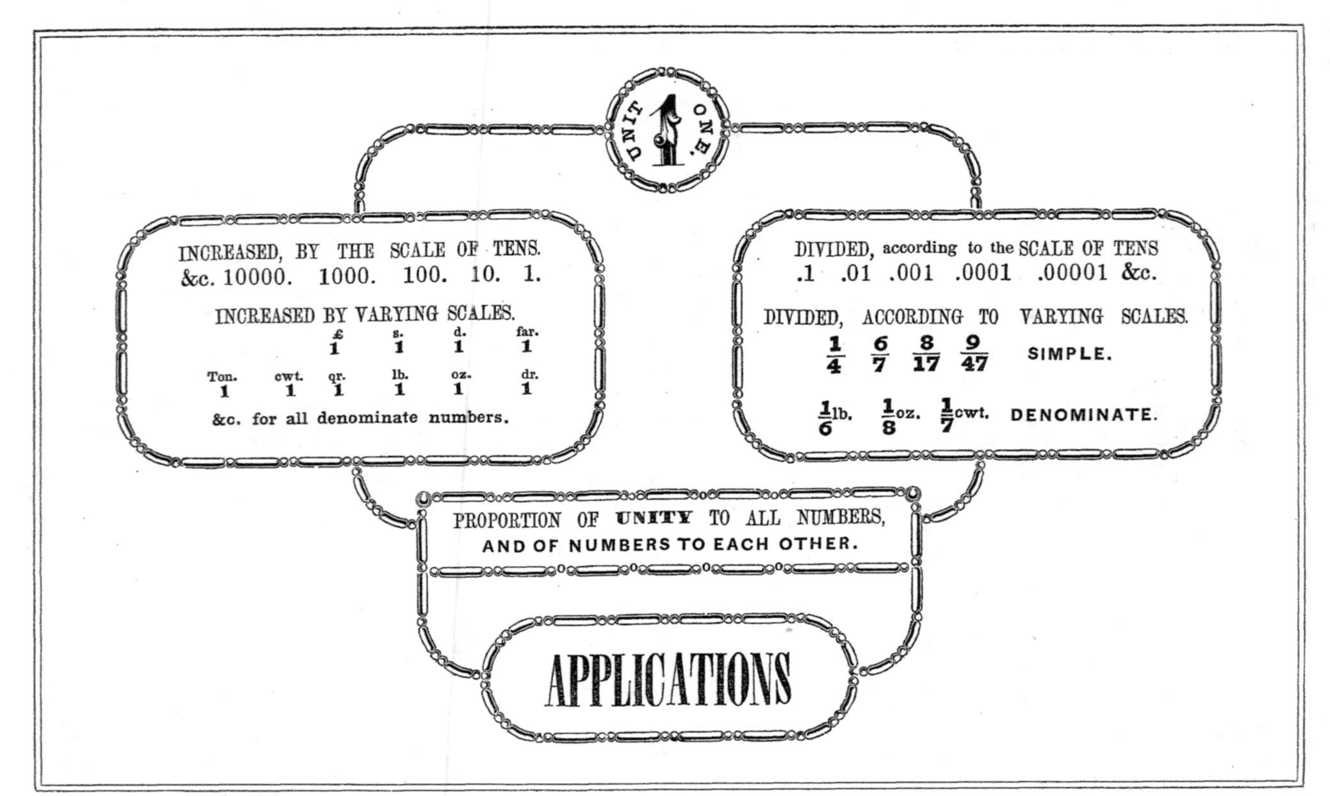

UNIT 1 ONE.
INCREASED, BY THE SCALE OF TENS.
&c. 10000. 1000. 100. 10. 1.
INCREASED BY VARYING SCALES.
£ 1 s. 1 d. 1 far. 1
Ton. 1 cwt. 1 qr. 1 lb. 1 oz. 1 dr. 1
&c. for all denominate numbers.
DIVIDED, according to the SCALE OF TENS
.1 .01 .001 .0001 .00001 &c.
DIVIDED, ACCORDING TO VARYING SCALES.
1/4 6/7 8/17 9/47 SIMPLE.
1/6 lb. 1/8 oz. 1/7 cwt. DENOMINATE.
PROPORTION OF UNITY TO ALL NUMBERS,
AND OF NUMBERS TO EACH OTHER.
APPLICATIONS

# CHAPTER II.

## ARITHMETIC—SCIENCE AND ART OF NUMBERS.

## SECTION I.

### INTEGER UNITS.

#### FIRST NOTIONS OF NUMBERS.

§ 104. THERE is but a single elementary idea in the science of numbers: it is the idea of the UNIT ONE. There is but one way of impressing this idea on the mind. It is by presenting to the senses a single object; as, one apple, one peach, one pear, &c.

*But one elementary idea in numbers.*

*How impressed on the mind.*

§ 105. There are three signs by means of which the idea of one is expressed and communicated. They are,

*Three signs for expressing it.*

1st. The word ONE. *A word.*

2d. The Roman character I. *Roman character:*

3d. The figure 1. *Figure.*

New ideas which arise by adding one.

§ 106. If one be added to one, the idea thus arising is different from the idea of one, and is complex. This new idea has also three signs; viz. TWO, II., and 2. If one be again added, that is, added to two, the new idea has likewise three signs; viz. THREE, III., and 3. The expressions for these, and similar ideas, are called numbers: hence,

The expressions are numbers.

Numbers defined.

NUMBERS *are expressions for one or more things of the same kind.*

### IDEAS OF NUMBERS GENERALIZED.

Ideas of numbers generalized.

§ 107. If we begin with the idea of the number one, and then add it to one, making two; and then add it to two, making three; and then to three, making four; and then to four, making five, and so on; it is plain that we shall form a series of numbers, each of which will be greater by one than that which precedes it. Now, one or unity, is the basis of this series of numbers, and each number may be expressed in three ways:

How formed.

Unity the basis.

Three ways of expressing them.

1st way. 1st. By the words ONE, TWO, THREE, &c., of our common language;

2d way. 2d. By the Roman characters; and,

3d way. 3d. By figures.

§ 108. Since all numbers, whether integer or fractional, must come from, and hence be connected with, the unit one, it follows that there is but one purely elementary idea in the science of numbers. Hence, the idea of every number, regarded as made up of units (and all numbers except one must be so regarded when we analyze them), is necessarily complex. For, since the number arises from the addition of ones, the apprehension of it is incomplete until we understand how those additions were made; and therefore, a full idea of the number is necessarily complex.

All numbers come from one:

Hence but one idea that is purely elementary.

All other notions are complex.

§ 109. But if we regard a number as an entirety, that is, as an entire or whole thing, as an entire two, or three, or four, without pausing to analyze the units of which it is made up, it may then be regarded as a simple or incomplex idea; though, as we have seen, such idea may always be traced to that of the unit one, which forms the basis of the number.

When a number may be regarded as incomplex.

### UNITY AND A UNIT DEFINED.

§ 110. When we name a number, as twenty feet, two things are necessary to its clear apprehension.

What is necessary to the apprehension of a number

First. 1st. A distinct apprehension of the *single thing* which forms the basis of the number; and,

Second. 2d. A distinct apprehension of the *number of times* which that thing is taken.

The basis of the number is UNITY. When it is called UNITY, and when a UNIT. The single thing, which forms the basis of the number, is called UNITY, or a UNIT. It is called unity, when it is regarded as the *primary basis* of the number; that is, when it is the final standard to which all the numbers that come from it are referred. It is called a unit when it is regarded as one of the collection of several equal things which form a number. Thus, in the example, one foot, regarded as a standard and the basis of the number, is called UNITY; but, considered as one of the twenty equal feet which make up the number, it is called a UNIT.

### OF SIMPLE AND DENOMINATE NUMBERS.

Abstract unit. § 111. A simple or abstract unit, is ONE, without regard to the kind of thing to which the term *one* may be applied.

Denominate unit. A denominate or concrete unit, is *one thing* named or denominated; as, one apple, one peach, one pear, one horse, &c.

Number has no reference § 112. Number, as such, has no reference to the particular things numbered. But to distinguish numbers which are applied to particular units from those which are purely abstract, we call the latter Abstract or Simple Numbers, and the former Concrete or Denominate Numbers. Thus, fifteen is an abstract or simple number, because the unit is *one;* and fifteen pounds is a concrete or denominate number, because its unit, one pound, is denominated or named. to the things numbered. Simple and Denominate. Examples.

### ALPHABET—WORDS—GRAMMAR.

§ 113. The term alphabet, in its most general sense, denotes a set of characters which form the elements of a written language. Alphabet.

When any one of these characters, or any combination of them, is used as the sign of a distinct notion or idea, it is called a word; and the naming of the characters of which the word is composed, is called its spelling. Words.

Grammar, as a science, treats of the established connection between words as the signs of ideas. Grammar.

### ARITHMETICAL ALPHABET.

§ 114. The arithmetical alphabet consists of ten characters, called figures. They are, Arithmetical Alphabet.

| Naught, | One, | Two, | Three, | Four, | Five, | Six, | Seven, | Eight, | Nine, |
|---|---|---|---|---|---|---|---|---|---|
| 0 | 1 | 2 | 3 | 4 | 5 | 6 | 7 | 8 | 9 |

and each may be regarded as a word, since it stands for a distinct idea.

WORDS—SPELLING AND READING IN ADDITION.

One cannot be spelled.

§ 115. The idea of one, being elementary, the character 1 which represents it, is also elementary, and hence, cannot be spelled by the other characters of the Arithmetical Alphabet (§ 114). But the idea which is expressed by 2 comes from the addition of 1 and 1: hence, the word represented by the character 2, may be spelled by 1 and 1. Thus, 1 and 1 are 2, is the arithmetical spelling of the word two.

Spelling by the arithmetical characters.

Three is spelled thus: 1 and 2 are 3; and also, 2 and 1 are 3.

Examples.

Four is spelled, 1 and 3 are 4; 3 and 1 are 4; 2 and 2 are 4.

Five is spelled, 1 and 4 are 5; 4 and 1 are 5; 2 and 3 are 5; 3 and 2 are 5.

Six is spelled, 1 and 5 are 6; 5 and 1 are 6; 2 and 4 are 6; 4 and 2 are 6; 3 and 3 are 6.

All numbers may be spelled in a similar way.

§ 116. In a similar manner, any number in arithmetic may be spelled; and hence we see that the process of spelling in addition consists simply, in naming any two elements which will make up the number. All the numbers in addition are therefore spelled with two syllables. The *reading* consists in naming only the word which expresses the final idea. Thus,

Reading: in what it consists.

Examples.

| 0 | 1 | 2 | 3 | 4 | 5 | 6 | 7 | 8 | 9 |
|---|---|---|---|---|---|---|---|---|---|
| 1 | 1 | 1 | 1 | 1 | 1 | 1 | 1 | 1 | 1 |
| One | two | three | four | five | six | seven | eight | nine | ten. |

We may now read the words which express the first hundred combinations.

READINGS.

| | | | | | | | | | | Read. |
|---|---|---|---|---|---|---|---|---|---|---|
| 1 | 2 | 3 | 4 | 5 | 6 | 7 | 8 | 9 | 10 | Two, three, four, &c. |
| 1 | 1 | 1 | 1 | 1 | 1 | 1 | 1 | 1 | 1 | |
| 1 | 2 | 3 | 4 | 5 | 6 | 7 | 8 | 9 | 10 | Three, four, &c. |
| 2 | 2 | 2 | 2 | 2 | 2 | 2 | 2 | 2 | 2 | |
| 1 | 2 | 3 | 4 | 5 | 6 | 7 | 8 | 9 | 10 | Four, five, &c. |
| 3 | 3 | 3 | 3 | 3 | 3 | 3 | 3 | 3 | 3 | |
| 1 | 2 | 3 | 4 | 5 | 6 | 7 | 8 | 9 | 10 | Five, six, &c. |
| 4 | 4 | 4 | 4 | 4 | 4 | 4 | 4 | 4 | 4 | |
| 1 | 2 | 3 | 4 | 5 | 6 | 7 | 8 | 9 | 10 | Six, seven, &c. |
| 5 | 5 | 5 | 5 | 5 | 5 | 5 | 5 | 5 | 5 | |
| 1 | 2 | 3 | 4 | 5 | 6 | 7 | 8 | 9 | 10 | Seven, eight, &c. |
| 6 | 6 | 6 | 6 | 6 | 6 | 6 | 6 | 6 | 6 | |
| 1 | 2 | 3 | 4 | 5 | 6 | 7 | 8 | 9 | 10 | Eight, nine, &c. |
| 7 | 7 | 7 | 7 | 7 | 7 | 7 | 7 | 7 | 7 | |
| 1 | 2 | 3 | 4 | 5 | 6 | 7 | 8 | 9 | 10 | Nine, ten, &c. |
| 8 | 8 | 8 | 8 | 8 | 8 | 8 | 8 | 8 | 8 | |

Ten, eleven, &c.

| 1 | 2 | 3 | 4 | 5 | 6 | 7 | 8 | 9 | 10 |
|---|---|---|---|---|---|---|---|---|---|
| 9 | 9 | 9 | 9 | 9 | 9 | 9 | 9 | 9 | 9 |

Eleven, twelve, &c.

| 1 | 2 | 3 | 4 | 5 | 6 | 7 | 8 | 9 | 10 |
|---|---|---|---|---|---|---|---|---|---|
| 10 | 10 | 10 | 10 | 10 | 10 | 10 | 10 | 10 | 10 |

Example for reading in Addition.

§ 117. In this example, beginning at the right hand, we say, 8, 17, 18, 26: setting down the 6 and carrying the 2, we say, 8, 13, 20, 22, 29: setting down the 9 and carrying the 2, we say, 9, 12, 18, 22, 30: and setting down the 30, we have the entire sum 3096. All the examples in addition may be done in a similar manner.

| |
|---:|
| 878 |
| 421 |
| 679 |
| 354 |
| 764 |
| 3096 |

All examples so solved.

Advantages of reading.

§ 118. The advantages of this method of reading over spelling are very great.

1st. stated.

1st. The mind acquires ideas more readily through the eye than through either of the other senses. Hence, if the mind be taught to apprehend the result of a combination, by merely seeing its elements, the process of arriving at it is much shorter than when those elements are presented through the instrumentality of sound. Thus, to *see* 4 and 4, and *think* 8, is a very different thing from saying, four and four are eight.

2d. stated.

2d. The mind operates with greater rapidity and certainty, the nearer it is brought to the ideas which it is to apprehend and combine. Therefore, all unnecessary words load it and impede its operations. Hence, to spell when we can read, is to fill the mind with words and sounds, instead of ideas.

3d. All the operations of arithmetic, beyond the elementary combinations, are performed on paper; and if rapidly and accurately done, must be done through the eye and by reading. Hence the great importance of beginning early with a method which must be acquired before any considerable skill can be attained in the use of figures. 3d. stated.

§ **119.** It must not be supposed that the *reading* can be accomplished until the *spelling* has first been learned. Reading comes after spelling.

In our common language, we first learn the alphabet, then we pronounce each letter in a word, and finally, we pronounce the word. We should do the same in the arithmetical reading. Same as in our common language.

### WORDS—SPELLING AND READING IN SUBTRACTION.

§ **120.** The processes of spelling and reading which we have explained in the addition of numbers, may, with slight modifications, be applied in subtraction. Thus, if we are to subtract Same principle applied in Subtraction.

2 from 5, we say, ordinarily, 2 from 5 leaves 3; or 2 from 5 three remains. Now, the word, three, is suggested by the relation in which 2 and 5 stand to each other, and this word may be read at once. Hence, *the reading, in subtraction, is simply naming the word, which expresses the difference between the subtrahend and minuend.* Thus, we may read each word of the following one hundred combinations.

Readings in Subtraction explained.

READINGS.

| | | | | | | | | | | |
|---|---|---|---|---|---|---|---|---|---|---|
| One from one, &c. | 1 | 2 | 3 | 4 | 5 | 6 | 7 | 8 | 9 | 10 |
| | 1 | 1 | 1 | 1 | 1 | 1 | 1 | 1 | 1 | 1 |
| Two from two, &c. | 2 | 3 | 4 | 5 | 6 | 7 | 8 | 9 | 10 | 11 |
| | 2 | 2 | 2 | 2 | 2 | 2 | 2 | 2 | 2 | 2 |
| Three from three, &c. | 3 | 4 | 5 | 6 | 7 | 8 | 9 | 10 | 11 | 12 |
| | 3 | 3 | 3 | 3 | 3 | 3 | 3 | 3 | 3 | 3 |
| Four from four, &c. | 4 | 5 | 6 | 7 | 8 | 9 | 10 | 11 | 12 | 13 |
| | 4 | 4 | 4 | 4 | 4 | 4 | 4 | 4 | 4 | 4 |
| Five from five, &c. | 5 | 6 | 7 | 8 | 9 | 10 | 11 | 12 | 13 | 14 |
| | 5 | 5 | 5 | 5 | 5 | 5 | 5 | 5 | 5 | 5 |
| Six from six, &c. | 6 | 7 | 8 | 9 | 10 | 11 | 12 | 13 | 14 | 15 |
| | 6 | 6 | 6 | 6 | 6 | 6 | 6 | 6 | 6 | 6 |
| Seven from seven, &c. | 7 | 8 | 9 | 10 | 11 | 12 | 13 | 14 | 15 | 16 |
| | 7 | 7 | 7 | 7 | 7 | 7 | 7 | 7 | 7 | 7 |

| 8 | 9 | 10 | 11 | 12 | 13 | 14 | 15 | 16 | 17 | Eight from eight, &c. |
|---|---|---|---|---|---|---|---|---|---|---|
| 8 | 8 | 8 | 8 | 8 | 8 | 8 | 8 | 8 | 8 | |

| 9 | 10 | 11 | 12 | 13 | 14 | 15 | 16 | 17 | 18 | Nine from nine, &c. |
|---|---|---|---|---|---|---|---|---|---|---|
| 9 | 9 | 9 | 9 | 9 | 9 | 9 | 9 | 9 | 9 | |

| 10 | 11 | 12 | 13 | 14 | 15 | 16 | 17 | 18 | 19 | Ten from ten, &c. |
|---|---|---|---|---|---|---|---|---|---|---|
| 10 | 10 | 10 | 10 | 10 | 10 | 10 | 10 | 10 | 10 | |

§ 121. It should be remarked, that in subtraction, as well as in addition, the spelling of the words must necessarily precede their reading. The spelling consists in naming the figures with which the operation is performed, the steps of the operation, and the final result. The reading consists in naming the final result only.

*Spelling precedes reading in Subtraction.*

*Reading.*

SPELLING AND READING IN MULTIPLICATION.

§ 122. Spelling in multiplication is similar to the corresponding process in addition or subtraction. It is simply naming the two elements which produce the product; whilst the reading consists in naming only the word which expresses the final result.

*Spelling in Multiplication.*

*Reading.*

In multiplying each number from 1 to 10 by 2, we usually say, two times 1 are 2; two times 2 are 4; two times 3 are 6; two times 4 are 8; two times 5 are 10; two times 6 are 12; two

*Examples in spelling.*

times 7 are 14; two times 8 are 16; two times 9 are 18; two times 10 are 20. Whereas, we should merely read, and say, 2, 4, 6, 8, 10, 12, 14, 16, 18, 20.

In reading.

In a similar manner we read the entire multiplication table.

READINGS.

| | | | | | | | | | | | | |
|---|---|---|---|---|---|---|---|---|---|---|---|---|
| Once one is 1, &c. | 12 | 11 | 10 | 9 | 8 | 7 | 6 | 5 | 4 | 3 | 2 | 1 |
| | | | | | | | | | | | | 1 |
| Two times 1 are 2, &c. | 12 | 11 | 10 | 9 | 8 | 7 | 6 | 5 | 4 | 3 | 2 | 1 |
| | | | | | | | | | | | | 2 |
| Three times 1 are 3, &c. | 12 | 11 | 10 | 9 | 8 | 7 | 6 | 5 | 4 | 3 | 2 | 1 |
| | | | | | | | | | | | | 3 |
| Four times 1 are 4, &c. | 12 | 11 | 10 | 9 | 8 | 7 | 6 | 5 | 4 | 3 | 2 | 1 |
| | | | | | | | | | | | | 4 |
| Five times 1 are 5, &c. | 12 | 11 | 10 | 9 | 8 | 7 | 6 | 5 | 4 | 3 | 2 | 1 |
| | | | | | | | | | | | | 5 |
| Six times 1 are six, &c. | 12 | 11 | 10 | 9 | 8 | 7 | 6 | 5 | 4 | 3 | 2 | 1 |
| | | | | | | | | | | | | 6 |
| Seven times 1 are 7, &c. | 12 | 11 | 10 | 9 | 8 | 7 | 6 | 5 | 4 | 3 | 2 | 1 |
| | | | | | | | | | | | | 7 |
| Eight times 1 are 8, &c. | 12 | 11 | 10 | 9 | 8 | 7 | 6 | 5 | 4 | 3 | 2 | 1 |
| | | | | | | | | | | | | 8 |

| 12 | 11 | 10 | 9 | 8 | 7 | 6 | 5 | 4 | 3 | 2 | 1 | Nine times 1 are 9, &c. |
|---|---|---|---|---|---|---|---|---|---|---|---|---|
| | | | | | | | | | | | 9 | |

| 12 | 11 | 10 | 9 | 8 | 7 | 6 | 5 | 4 | 3 | 2 | 1 | Ten times 1 are 10, &c. |
|---|---|---|---|---|---|---|---|---|---|---|---|---|
| | | | | | | | | | | | 10 | |

| 12 | 11 | 10 | 9 | 8 | 7 | 6 | 5 | 4 | 3 | 2 | 1 | Eleven times 1 are 11, &c. |
|---|---|---|---|---|---|---|---|---|---|---|---|---|
| | | | | | | | | | | | 11 | |

| 12 | 11 | 10 | 9 | 8 | 7 | 6 | 5 | 4 | 3 | 2 | 1 | Twelve times 1 are 12, &c. |
|---|---|---|---|---|---|---|---|---|---|---|---|---|
| | | | | | | | | | | | 12 | |

## SPELLING AND READING IN DIVISION.

§ 123. In all the cases of short division, the quotient may be read immediately without naming the process by which it is obtained. Thus, in dividing the following numbers by 2, we merely read the words below. *In Short Division, we may read:*

| 2)4 | 6 | 8 | 10 | 12 | 16 | 18 | 22 |
|---|---|---|---|---|---|---|---|
| two | three | four | five | six | eight | nine | eleven. |

In a similar manner, all the words, expressing the results in short division, may be read. *In all cases.*

### READINGS.

| 2)2 | 4 | 6 | 8 | 10 | 12 | 14 | 16 | 18 | 20 | 22 | 24 | Two in 2, once, &c. |
|---|---|---|---|---|---|---|---|---|---|---|---|---|
| 3)3 | 6 | 9 | 12 | 15 | 18 | 21 | 24 | 27 | 30 | 33 | 36 | Three in 3, once, &c. |
| 4)4 | 8 | 12 | 16 | 20 | 24 | 28 | 32 | 36 | 40 | 44 | 48 | Four in 4, once, &c. |

| | |
|---|---|
| Five in 5, once, &c. | 5)5 10 15 20 25 30 35 40 45 50 55 60 |
| Six in 6, once, &c. | 6)6 12 18 24 30 36 42 48 54 60 66 72 |
| Seven in 7, once, &c. | 7)7 14 21 28 35 42 49 56 63 70 77 84 |
| Eight in 8, once, &c. | 8)8 16 24 32 40 48 56 64 72 80 88 96 |
| Nine in 9, once, &c. | 9) 9 18 27 36 45 54 63 72 81 90 99 108 |
| Ten in 10, once, &c. | 10)10 20 30 40 50 60 70 80 90 100 110 120 |
| Eleven in 11, once, &c. | 11)11 22 33 44 55 66 77 88 99 110 121 132 |
| Twelve in 12, once, &c. | 12)12 24 36 48 60 72 84 96 108 120 132 144 |

UNITS INCREASING BY THE SCALE OF TENS.

The idea of a particular number is complex.

§ 124. The idea of a particular number is necessarily complex; for, the mind naturally asks:

1st. What is the unit or basis of the number? and,

2d. How many times is the unit or basis taken?

What a figure indicates.

§ 125. A figure indicates how many times a unit is taken. Each of the ten figures, however written, or however placed, always expresses as many units as its name imports, and no more; nor does the *figure itself* at all indicate the kind

Number has one for its basis.

of unit. Still, every number expressed by one or more figures, has for its basis either the abstract unit one, or a denominate unit.* If a denominate unit, its value or kind is pointed out either by our common language, or as we shall presently see, by the *place* where the figure is written.

Number expressed by a single figure.

The *number* of units which may be expressed by either of the ten figures, is indicated by the name of the figure. If the figure stands alone, and the unit is not denominated, the basis of the number is the abstract unit 1.

How ten is written.

§ 126. If we write 0 on the right of 1, we have - - - - - - - - - - 10,

which is read ONE ten. Here 1 still expresses ONE, but it is ONE ten; that is, a unit ten times as great as the unit 1; and this is called a unit of the *second order*.

Unit of the second order.

How to write one hundred.

Again; if we write two 0's on the right of 1, we have - - - - - - - 100,

which is read ONE hundred. Here again, 1 still expresses ONE, but it is ONE hundred; that is, a unit ten times as great as the unit ONE ten, and a hundred times as great as the unit 1.

A unit of the third order.

Laws—when figures are written by the side of each other.

§ 127. If three 1's are written by the side of each other, thus - - - - 111,

* Section 111.

the ideas, expressed in our common language, are these:

First.

1st. *That the* 1 *on the right, will either express a single thing denominated, or the abstract unit one.*

Second.

2d. *That the* 1 *next to the left expresses* 1 *ten; that is, a unit ten times as great as the first.*

Third.

3d. *That the* 1 *still further to the left expresses* 1 *hundred; that is, a unit ten times as great as the second, and one hundred times as great as the first; and similarly if there were other places.*

What the language establishes when figures are so written.

When figures are thus written by the side of each other, the arithmetical language establishes a relation between the units of their places: that is, the unit of each place, as we pass from the right hand towards the left, increases according to the scale of tens. Therefore, by a law of the arithmetical language, *the place of a figure fixes its unit.*

Scale for Numeration.

If, then, we write a row of 0's as a scale, thus:

| 1 hundred billion, | 1 ten billion, | 1 billion, | 1 hundred million, | 1 ten million, | 1 million, | 1 hundred thousand, | 1 ten thousand, | 1 thousand, | 1 hundred, | 1 ten, | 1 unit, |
|---|---|---|---|---|---|---|---|---|---|---|---|
| 0 | 0 | 0, | 0 | 0 | 0, | 0 | 0 | 0, | 0 | 0 | 0 |

The units of place determined.

the *unit of each place* is determined, as well

as the *law* of change in passing from one place to another. If then, it were required to express a given number of units, of any order, we first select from the arithmetical alphabet the character which designates the number, and then write it in the place corresponding to the order. Thus, to express three millions, we write

How any number of units may be expressed.

**3000000**;

and similarly for all numbers.

§ **128**. It should be observed, that a figure being *a character which represents value*, can have no value in and of itself. The *number* of things, which any figure expresses, is determined by its name, as given in the arithmetical alphabet. The *kind* of thing, or unit of the figure, is fixed either by naming it, as in the case of a denominate number, or by the place which the figure occupies, when written by the side of or over* other figures.

A figure has no value in itself.

How the unit is determined.

The phrase "local value of a figure," so long in use, is, therefore, without signification when applied to a figure: the term "local value," being applicable to the *unit of the place*, and not to the figure which occupies the place.

Figure, has no local value.

Term applicable to *unit of place*.

§ **129**. Federal Money affords an example of a

Federal Money.

* Section 199.

Its denominations. series of denominate units, increasing according to the scale of tens: thus,

| Eagle, | Dollar, | Dime, | Cent, | Mill, |
|---|---|---|---|---|
| 1 | 1 | 1 | 1 | 1 |

How read. may be read 11 thousand 1 hundred and 11 *mills;* or, 1111 *cents* and 1 mill; or, 111 dimes 1 cent and 1 mill; or, 11 dollars 1 dime 1 cent and 1 mill; or, 1 eagle 1 dollar 1 dime 1 cent and 1 mill. Various kinds of Readings. Thus, we may read the number with either of its units as a basis, or we may name them all: thus, 1 eagle, 1 dollar, 1 dime, 1 cent, 1 mill. Generally, in Federal Money, we read in the denominations of dollars, cents, and mills; and should say, 11 dollars 11 cents and 1 mill.

Examples in Reading. § 130. Examples in reading figures:—

1st. Example. If we have the figures - - - - 89
we may read them by their smallest unit, and say eighty-nine; or, we may say 8 tens and 9 units.

2d. Example. Again, the figures - - - - - - - 567
may be read by the smallest unit; viz. five hundred and sixty-seven; or we may say, 56 tens and 7 units; or, 5 hundreds 6 tens and 7 units.

3d. Example. Again, the number expressed by - 74896

may be read, seventy-four thousand eight hundred and ninety-six. Or, it may be read, 7489 tens and 6 units; or, 748 hundreds 9 tens and 6 units; or, 74 thousands 8 hundreds 9 tens and 6 units; or, 7 ten thousands 4 thousands 8 hundreds 9 tens and 6 units; and we may read in a similar way all other numbers.

Various readings of a number.

Although we should teach all the correct readings of a number, we should not fail to remark that it is generally most convenient in practice to read by the lowest unit of a number. Thus, in the numeration table, we read each period by the lowest unit of that period. For example, in the number

The best method of reading.

Each period read by its lowest unit.

874,967,847,047,

Example.

we read 874 *billions* 967 *millions* 847 *thousands* and 47.

### UNITS INCREASING ACCORDING TO VARYING SCALES.

§ 131. If we write the well-known signs of the English money, and place 1 under each denomination, we shall have

Methods of writing figures having different denominate units.

| £. | *s.* | *d.* | *f.* |
|---|---|---|---|
| 1 | 1 | 1 | 1 |

Now, the signs £. *s.* *d.* and *f.* fix the value of the unit 1 in each denomination; and they also

How the value of each unit is fixed.

What the language expresses.

determine the relations which subsist between the different units. For example, this simple language expresses these ideas:

The units of the places.

1st. That the unit of the right-hand place is 1 farthing—of the place next to the left, 1 penny—of the next place, 1 shilling—of the next place, 1 pound; and

How the units increase.

2d. That 4 units of the lowest denomination make one unit of the next higher; 12 of the second, one of the third; and 20 of the third, one of the fourth.

The units in Avoirdupois weight.

If we take the denominate numbers of the Avoirdupois weight, we have

| *Ton.* | *cwt.* | *qr.* | *lb.* | *oz.* | *dr.* |
|---|---|---|---|---|---|
| 1 | 1 | 1 | 1 | 1 | 1; |

Changes in the value of the units.

in which the units increase in the following manner: viz. the second unit, counting from the right, is sixteen times as great as the first; the third, sixteen times as great as the second; the fourth, twenty-five times as great as the third; the fifth, four times as great as the fourth; and the sixth, twenty times as great as the fifth.

How the scale varies.

The scale, therefore, for this class of denominate numbers varies according to the above laws.

A different scale for each system.

If we take any other class of denominate numbers, as the Troy weight, or any of the systems of measures, we shall have different scales for the formation of the different units.

But in all the formations, we shall recognise the application of the same general principles.

*The method of forming the scales the same for all numbers.*

There are, therefore, two general methods of forming the different systems of integer numbers from the unit one. The first consists in preserving a constant law of relation between the different unities; viz. that their values shall change according to the scale of tens. This gives the system of common numbers.

*Two systems of forming integer numbers.*

*First system.*

The second method consists in the application of known, though varying laws of change in the unities. These changes in the unities produce the entire system of denominate numbers, each class of which has its appropriate scale, and the changes among the units of the same class are indicated by the different degrees of its scale.

*Second system.*

*Change in the scales of forming the unities.*

### INTEGER UNITS OF ARITHMETIC.

§ 132. There are four principal classes of units in arithmetic:

*Four classes of units.*

1st. Abstract, or simple units; *1st. class.*

2d. Units of Currency; *2d. class.*

3d. Units of Weight; and *3d. class.*

4th. Units of Measure. *4th. class.*

First among the Units of arithmetic stands the simple or abstract unit 1. This is the basis of all simple numbers, and becomes the basis,

*Abstract unit one, the basis.*

The basis of denominate numbers; also, of all denominate numbers, by merely naming, in succession, the particular things to which it is applied.

Also, the basis of all fractions, whether simple or denominate. It is also the basis of all fractions. Merely as the unit 1, it is a whole which may be divided according to any law, forming every variety of fraction; and if we apply it to a particular thing, the fraction becomes denominate, and we have expressions for all conceivable parts of that thing.

§ 133. It has been remarked* that we can form no distinct apprehension of a number, until we have a clear notion of its unit, and the number of times the unit is taken. *The unit is the great basis.* The utmost care, therefore, should be taken to impress on the minds of learners, a clear and distinct idea of the actual value of the unit of every number with which they have to do. If it be a number expressing currency, one or more of the coins should be exhibited, and the value dwelt upon; after which, distinct notions of the other units can be acquired by comparison.

Must apprehend the unit.

Let its nature and kind be fully explained;

How for a number expressing currency.

If the number be one of weight, some unit should be exhibited, as one pound, or one ounce, and an idea of its weight acquired by actually

Exhibit the unit if it be of weight;

* Section 110.

lifting it. This is the only way in which we can learn the true signification of the terms.

If the number be one of measure, either linear, superficial, liquid, or solid, its unit should also be exhibited, and *the signification of the term expressing it, learned in the only way in which it can be learned, through the senses, and by the aid of a sensible object.*

And also, if it be one of measure.

## FEDERAL MONEY.

§ 134. The currency of the United States is called Federal Money. Its units are all denominate, being 1 mill, 1 cent, 1 dime, 1 dollar, 1 eagle. The law of change, in passing from one unit to another, is according to the scale of tens. Hence, this system of numbers may be treated, in all respects, as simple numbers; and indeed they are such, with the single exception that their units have different names.

Currency of the U. States.

Law of change in the unities.

How these numbers may be treated.

They are generally read in the units of dollars, cents, and mills—a comma being placed after the figure denoting dollars. Thus,

How generally read.

$$\$\,864{,}849$$

Example.

is read eight hundred and sixty-four dollars, eighty-four cents, and nine mills; and if there were a figure after the 9, it would be read in decimals of the mill. The number may, how-

Of figures after mills.

The number read in various ways. ever, be read in any other unit; as, 864849 mills; or, 86484 cents and 9 mills; or, 8648 dimes, 4 cents, and 9 mills; or, 86 eagles, 4 dollars, 84 cents, and 9 mills; and there are yet several other readings.

## ENGLISH MONEY.

Sterling Money. § 135. The units of English, or Sterling Money, are 1 farthing, 1 penny, 1 shilling, and 1 pound.

Scale of the unities. The scale of this class of numbers is a varying scale. Its degrees, in passing from the unit of the lowest denomination to the highest, are four, How it changes. twelve, and twenty. For, four farthings make one penny, twelve pence one shilling, and twenty shillings one pound.

## AVOIRDUPOIS WEIGHT.

Units in Avoirdupois. § 136. The units of the Avoirdupois Weight are 1 dram, 1 ounce, 1 pound, 1 quarter, 1 hundred-weight, and 1 ton.

Scale. The scale of this class of numbers is a varying scale. Its degrees, in passing from the unit of the lowest denomination to the highest, are sixteen, sixteen, twenty-five, four, and twenty. Variation in its degrees. For, sixteen drams make one ounce, sixteen ounces one pound, twenty-five pounds one quarter, four quarters one hundred, and twenty hundreds one ton.

### TROY WEIGHT.

§ 137. The units of the Troy Weight are, 1 grain, 1 pennyweight, 1 ounce, and 1 pound. Units in Troy Weight.

The scale is a varying scale, and its degrees, in passing from the unit of the lowest denomination to the highest, are twenty-four, twenty, and twelve. Scale: Its degrees.

### APOTHECARIES' WEIGHT.

§ 138. The units of this weight are, 1 grain, 1 scruple, 1 dram, 1 ounce, and 1 pound. Units in Apothecaries Weight.

The scale is a varying scale. Its degrees, in passing from the unit of the lowest denomination to the highest, are twenty, three, eight, and twelve. Scale: Its degrees.

### UNITS OF MEASURE.

§ 139. There are three units of measure, each differing in *kind* from the other two. They are, Units of Length, Units of Surface, and Units of Solidity. Three units of measure.

### UNITS OF LENGTH.

§ 140. The unit of length is used for measuring lines, either straight or curved. It is a Units of length.

The standard. straight line of a given length, and is often called the standard of the measurement.

What units are taken. The units of length, generally used as standards, are 1 inch, 1 foot, 1 yard, 1 rod, 1 furlong, and 1 mile. The number of times which the unit, used as a standard, is taken, considered in connection with its value, gives the idea of the length of the line measured. (Idea of length.)

## UNITS OF SURFACE.

Units of surface. § 141. Units of surface are used for the measurement of the area or contents of whatever has the two dimensions of length and breadth. The unit of surface is a square described on the unit of length as a side. Thus, if the unit of length be 1 foot, the corresponding unit of surface will be 1 square foot; that is, a square constructed on 1 foot of length as a side. (What the unit of surface is. Examples.)

1 square foot.

Its connection with the unit of length. If the linear unit be 1 yard, the corresponding unit of surface will be 1 square yard. It will be seen from the figure, that, although the linear yard contains the linear foot but three times, the square yard (Square feet in a square yard.)

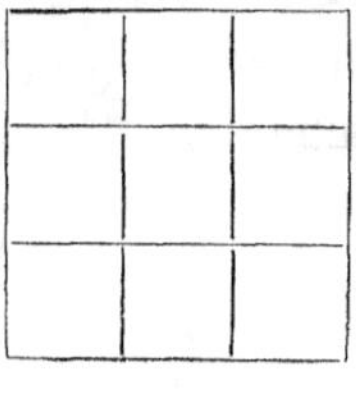

1 yard.

contains the square foot nine times. The square rod or square mile may also be used as the unit of surface. *Square rod and Square mile.*

The number of times which a surface contains its unit of measure, is its area or contents ; and this number, taken in connection with the value of the unit, gives the idea of its extent. *Area or contents of a surface.*

Besides the units of surface already considered, there is another kind, called,

## DUODECIMAL UNITS.

§ **142.** The duodecimal units are generally used in board measure, though they may be used in all superficial measurements, and also in solid. *Duodecimal units.*

The square foot is the basis of this class of units, and the others are deduced from it, by a descending scale of twelve. *Their basis.*

§ **143.** It is proved in Geometry, that if the number of linear units in the base of a rectangle be multiplied by the number of linear units in the height, the numerical value of the product will be equal to the number of superficial units in the figure. *What principle is proved in Geometry.*

Knowing this fact, we often express it by saying, that "feet multiplied by feet give square feet," and "yards multiplied by yards give square *How it is expressed.*

This a concise expression.

yards." But as feet cannot be taken *feet times*, nor yards *yard times*, this language, rightly understood, is but a concise form of expression for the principle stated above.

Conclusion.

With this understanding of the language, we say, that 1 foot in length multiplied by 1 foot in height, gives a square foot; and 4 feet in length multiplied by 3 feet in height, gives 12 square feet.

Examples in the multiplication of feet by feet and inches.

§ 144. If now, 1 foot in length be multiplied by 1 inch $=\frac{1}{12}$ of a foot in height, the product will be one-twelfth of a square foot; that is, *one-twelfth of the first unit:* if it be multiplied by 3 inches, the product will be three-twelfths of a square foot; and similarly for a multiplier of any number of inches.

Generalization.

Inches by inches.

If, now, we multiply 1 inch by 1 inch, the product may be represented by 1 square inch: that is, *by one-twelfth of the last unit.* Hence, *the units of this measure decrease according to the scale of* 12. The units are,

How the units change, and what they are.

First.

1st. Square feet—arising from multiplying feet by feet.

Second.

2d. Twelfths of square feet—arising from multiplying feet by inches.

3d. Twelfths of twelfths—arising from multiplying inches by inches. Third.

The same remarks apply to the smaller divisions of the foot, according to the scale of twelve. Conclusion general.

The difficulty of computing in this measure arises from the changes in the units. Difficulty.

### UNITS OF SOLIDITY.

§ 145. It has already been stated, that if length be multiplied by breadth, the product may be represented by units of surface. It is also proved, in Geometry, that if the length, breadth, and height of any regular solid body, of a square form, be multiplied together, the product may be represented by solid units whose number is equal to this product. Each solid unit is a cube constructed on the linear unit as an edge. Thus, if the linear unit be 1 foot, the solid unit will be 1 cubic or solid foot; that is, a cube constructed on 1 foot as an edge; and if it be 1 yard, the unit will be 1 solid yard. Units of solidity. What is proved in Geometry in regard to them. Solid units. Examples.

The three units, viz. the unit of length, the unit of surface, and the unit of solidity, are essentially different in kind. The first is a line of a known length; the second, a square of a known side: and the third, a solid, called a The three units essentially different. What they are.

Generally used. cube, of a known base and height. These are the units used in all kinds of measurement—excepting only the duodecimal system, which has already been explained. Duodecimal system.

### LIQUID MEASURE.

Units of Liquid Measure. § 146. The units of Liquid Measure are, 1 gill, 1 pint, 1 quart, 1 gallon, 1 barrel, 1 hogshead, 1 pipe, 1 tun. The scale is a varying scale. Its degrees, in passing from the unit of the lowest denomination, are, four, two, four, thirty-one and a half, sixty-three, two, and two.

Scale.

How it varies.

### DRY MEASURE.

Units of Dry Measure. § 147. The units of this measure are, 1 pint, 1 quart, 1 peck, 1 bushel, and 1 chaldron. The degrees of the scale, in passing from units of the lowest denomination, are two, eight, four, and thirty-six.

Degrees of the scale.

### TIME.

Units of Time. § 148. The units of Time are, 1 second, 1 minute, 1 hour, 1 day, 1 week, 1 month, 1 year, and 1 century. The degrees of the scale, in passing from units of the lowest denomination to the highest, are sixty, sixty, twenty-four, seven, four, twelve, and one hundred.

Degrees of the scale.

## CIRCULAR MEASURE.

§ 149. The units of this measure are, 1 second, 1 minute, 1 degree, 1 sign, 1 circle. The degrees of the scale, in passing from units of the lowest denomination to those of the higher, are sixty, sixty, thirty, and twelve. *Units of Circular Measure. Degrees of the Scale.*

## ADVANTAGES OF THE SYSTEM OF UNITIES.

§ 150. It may well be asked, if the method here adopted, of presenting the elementary principles of arithmetic, has any advantages over those now in general use. It is supposed to possess the following: *Advantages of the system.*

1st. The system of unities teaches an exact analysis of all numbers, and unfolds to the mind the different ways in which they are formed from the unit one, as a basis. *1st. Teaches the analysis of numbers:*

2d. Such an analysis enables the mind to form a definite and distinct idea of every number, by pointing out the relation between it and the unit from which it was derived. *2d. Points out their relation:*

3d. By presenting constantly to the mind the idea of the unit one, as the basis of all numbers, the mind is insensibly led to compare this unit with all the numbers which flow from it, and *3d. Constantly presents the idea of unity.*

then it can the more easily compare those numbers with each other.

4th. Explains more fully the four ground rules.

4th. It affords a more satisfactory analysis, and a better understanding of the four ground rules, and indeed of all the operations of arithmetic, than any other method of presenting the subject.

## FOUR GROUND RULES.

System applied in addition.

§ 151. Let us take the two following examples in Addition, the one in simple and the other in denominate numbers, and then analyze the process of finding the sum in each.

Examples.

| SIMPLE NUMBERS. | DENOMINATE NUMBERS | | | | |
|---|---|---|---|---|---|
| | *cwt.* | *qr.* | *lb.* | *oz.* | *dr.* |
| 874198 | 3 | 3 | 24 | 15 | 14 |
| 36984 | 6 | 3 | 23 | 14 | 8 |
| 3641 | | | | | |
| 914823 | 10 | 3 | 23 | 14 | 6 |

Process of performing addition.

In both examples we begin by adding the units of the lowest denomination, and then, we *divide their sum by so many as make one of the denomination next higher.* We then set down the remainder, and add the quotient to the units of that denomination. Having done this, we apply a similar process to all the other denominations—*the principle being precisely the same in both examples.* We see, in these examples, an

But one principle.

illustration of a general principle of addition, viz. *that units of the same kind are always added together.* Units of the same kind unite.

§ 152. Let us take two similar examples in Subtraction. System applied in subtraction.

| SIMPLE NUMBERS. |
| --- |
| 8403 |
| 3298 |
| 5105 |

| DENOMINATE NUMBERS. | | | |
| --- | --- | --- | --- |
| £ | *s.* | *d.* | *far.* |
| 6 | 9 | 7 | 2 |
| 3 | 10 | 8 | 4 |
| 2 | 18 | 10 | 2 |

Examples.

In both examples we begin with the units of the lowest denomination, and as the number in the subtrahend is greater than in the place directly above, we suppose so many to be added in the minuend as make one unit of the next higher denomination. We then make the subtraction, and add 1 to the units of the subtrahend next higher, and proceed in a similar manner, through all the denominations. It is plain that the principle employed is the same in both examples. Also, that units of any denomination in the subtrahend are taken from those of the same denomination in the minuend. The method of performing the examples. Principle the same for all examples.

§ 153. Let us now take similar examples in Multiplication. Multiplication.

Examples.

| SIMPLE NUMBERS. |
|---:|
| 87464 |
| 5 |
| 437320 |

| DENOMINATE NUMBERS. | | | | |
|---:|---:|---:|---:|---:|
| ℔ | ℥ | ʒ | ℈ | *gr.* |
| 9 | 7 | 6 | 2 | 15 |
| | | | | 5 |
| 48 | 3 | 2 | 1 | 15 |

Method of performing the examples.

In these examples we see, that we multiply, in succession, each order of units in the multiplicand by the multiplier, and that we carry from one product to another, one for every so many as make one unit of the next higher denomination. The *principle* of the process is therefore the same in both examples.

The principle the same for all examples.

§ 154. Finally, let us take two similar examples in Division.

Division.

Examples.

| SIMPLE NUMBERS. |
|---:|
| 3)874911 |
| 291637 |

| DENOMINATE NUMBERS. | | | |
|---:|---:|---:|---:|
| £ | *s.* | *d.* | *far.* |
| 3)8 | 4 | 2 | 1 |
| 2 | 14 | 8 | 3 |

Principles involved:

We begin, in both examples, by dividing the units of the highest denomination. The unit of the quotient figure is the same as that of the dividend. We write this figure in its place, and then reduce the remainder to units of the next lower denomination. We then add in that denomination, and continue the division through all the denominations to the last—the principle being precisely the same in both examples.

The same as in the other rules.

## SECTION II.

### FRACTIONAL UNITS.

---

#### FRACTIONAL UNITS.—SCALE OF TENS.

§ 155. If the unit 1 be divided into ten equal parts, each part is called *one tenth.* If one of these tenths be divided into ten equal parts, each part is called *one hundredth.* If one of the hundredths be divided into ten equal parts, each part is called *one thousandth;* and corresponding names are given to similar parts, how far soever the divisions may be carried.

Fraction one-tenth defined; One hundredth; One thousandth. Generalization.

Now, although the tenths which arise from dividing the unit 1, are but equal parts of 1, they are, nevertheless, WHOLE tenths, and in this light may be regarded as *units.*

Fractions are whole things.

To avoid confusion, in the use of terms, we shall call every equal part of 1 a *fractional unit.* Hence, tenths, hundredths, thousandths, tenths of thousandths, &c., *are fractional units, each having a fixed relation to the unit* 1, from which it was derived.

Fractional units. Their nature.

Fractional units of the first order; second order, &c.

§ 156. Adopting a similar language to that used in integer numbers, we call the tenths, fractional units of the *first order;* the hundredths, fractional units of the *second order;* the thousandths, fractional units of the *third order;* and so on for the subsequent divisions.

Language for fractional units.

Is there any arithmetical language by which these fractional units may be expressed? The decimal point, which is merely a dot, or period, indicates the division of the unit 1, according to the scale of tens. By the arithmetical language, the unit of the place next the point, on the right, is 1 tenth; that of the second place, 1 hundredth; that of the third, 1 thousandth; that of the fourth, 1 ten thousandth; and so on for places still to the right.

What it fixes.

Names of the places.

Scale.

The scale for decimals, therefore, is

.000000000, &c.;

in which the unit of each place is known as soon as we have learned the signification of the language.

If, therefore, we wish to express any of the parts into which the unit 1 may be divided, according to the scale of tens, we have simply to select from the alphabet, the figure that will express the *number* of parts, and then write it in

Any decimal number may be expressed by this scale.

the place corresponding to the *order of the unit.* Thus, to express four tenths, three thousandths, eight ten-thousandths, and six millionths, we write

Where any figure is written.

.403806 ;

Example.

and similarly, for any decimal which can be named.

§ 157. It should be observed that while the units of place *decrease*, according to the scale of tens, from left to right, they *increase* according to the same scale, from right to left. *This is the same law of increase as that which connects the units of place in simple numbers.* Hence, simple numbers and decimals being formed according to the same law, may be written by the side of each other and treated as a single number, by merely preserving the separating or decimal point. Thus, 8974 and .67046 may be written

The units increase from right to left.

Consequence.

8974.67046 ;

Example.

since ten units, in the place of tenths, make the unit one in the place next to the left.

## FRACTIONAL UNITS IN GENERAL.

§ 158. If the unit 1 be divided into two equal parts, each part is called a half. If it be divided

A half.

A third. into three equal parts, each part is called a third: if it be divided into four equal parts, each part is called a fourth: if into five equal parts, each part is called a fifth; and if into any number of equal parts, a name is given corresponding to the number of parts.

A fourth. A fifth. Generally.

These units are whole things. Now, although these halves, thirds, fourths, fifths, &c., are each but parts of the unit 1, they are, nevertheless, in *themselves*, whole things. That is, a half is a whole half; a third, a whole third; a fourth, a whole fourth; and the same for any other equal part of 1. In this sense, therefore, they are *units*, and we call them fractional units. Each is an exact part of the unit 1, and has a fixed relation to it.

Examples. Have a relation to unity.

§ 159. Is there any arithmetical language by which these fractional units can be expressed?

Language for fractions. The bar, written at the right, is the sign which denotes the division of the unit 1 into any number of equal parts.

$$\overline{\phantom{9}}$$

To express the number of equal parts. If we wish to express the number of equal parts into which it is divided, as 9, for example, we simply write the 9 under the bar, and then the phrase means, that some thing regarded as a whole, has been divided into 9 equal parts.

$$\overline{9}$$

If, now, we wish to express any number of these fractional units, as 7, for example, we place the 7 above the line, and read, seven ninths.

$$\frac{7}{9}$$

To show how many are taken.

§ 160. It was observed,* that two things are necessary to the clear apprehension of an integer number.

Two things necessary to apprehend a number.

1st. A distinct apprehension of the *unit* which forms the basis of the number; and, — First.

2dly. A distinct apprehension of the number of times which that unit is taken. — Second.

Three things are necessary to the distinct apprehension of the value of any fraction, either decimal or vulgar.

Three things necessary to apprehend a fraction.

1st. We must know the unit, or whole thing, from which the fraction was derived; — First.

2d. We must know into how many equal parts that unit is divided; and, — Second.

3dly. We must know how many such parts are taken in the expression. — Third.

The unit from which the fraction is derived, is called the *unit of the fraction;* and one of the equal parts is called, the *unit of the expression.*

Unit of the fraction—of the expression.

For example, to apprehend the value of the

---

* Section 110.

What we must know. fraction $\frac{3}{7}$ of a pound avoirdupois, or $\frac{3}{7}$*lb.*; we must know,

First. 1st. What is meant by a pound;

Second. 2d. That it has been divided into seven equal parts; and,

Third. 3d. That three of those parts are taken.

In the above fraction, 1 pound is the unit of the fraction; one-seventh of a pound, the unit of the expression; and **3** denotes that three fractional units are taken.

Unit when not named. If the unit of a fraction be not named, it is taken to be the abstract unit **1**.

## ADVANTAGES OF FRACTIONAL UNITS.

Every equal part of one, a unit. § **161.** By considering every equal part of unity as a unit of itself, having a certain relation to the unit 1, the mind is led to analyze a fraction, and thus to apprehend its precise signification.

Advantages of the analysis. Under this searching analysis, the mind at once seizes on the *unit of the fraction* as the principal basis. It then looks at the value of each part. It then inquires how many such parts are taken.

Equal units, whether integral or frac- It having been shown that equal integer units can alone be added, it is readily seen that the same principle is equally applicable to fractional units; and then the inquiry is made: What is necessary in order to make such units equal?

tional, can alone be added.

It is seen at once, that two things are necessary:

Two things necessary for addition.

1st. That they be parts of the *same unit;* and,

First.

2d. That they be *like parts;* in other words, they must be of the same denomination, and have a common denominator.

Second.

In regard to Decimal Fractions, all that is necessary, is to observe that units of the same value are added to each other, and when the figures expressing them are written down, they should always be placed in the same column.

Decimal Fractions.

§ 162. The great difficulty in the management of fractions, consists in comparing them with each other, instead of constantly comparing them with the unity from which they are derived. By considering them as entire things, having a fixed relation to the unity which is their basis, they can be compared as readily as integer numbers; for, the mind is never at a loss when it apprehends the unit, the parts into which it is divided, and the number of parts which are taken. The only reasons why we apprehend and

Difficulty in the management of fractions.

How obviated.

Reasons for greater simplicity in integers.

handle integer numbers more readily than fractions, are,

First. 1st. Because the unity forming the basis is always kept in view; and,

Second. 2d. Because, in integer numbers, we have been taught to trace constantly the connection between the unity and the numbers which come from it; while in the methods of treating fractions, these important considerations have been neglected.

---

## SECTION III.

### PROPORTION AND RATIO.

Proportion defined. § 163. PROPORTION expresses the relation which one number bears to another, with respect to its being greater or less.

Two ways of comparing. Two numbers may be compared, the one with the other, in two ways:

1st method. 1st. With respect to their difference, called Arithmetical Proportion; and,

2d method. 2d. With respect to their quotient, called Geometrical Proportion.

Thus, if we compare the numbers 1 and 8, by their difference, we find that the second exceeds the first by 7: hence, their difference 7, *is the measure of their arithmetical proportion,* and is called, in the old books, their *arithmetical ratio.*

Example of Arithmetical Proportion.

Arithmetical Ratio.

If we compare the same numbers by their quotient, we find that the second contains the first 8 times: hence, 8 is the *measure of their geometrical proportion,* and is called their *geometrical ratio.**

Example of Geometrical Proportion.

Ratio.

§ 164. The two numbers which are thus compared, are called *terms.* The first is called the *antecedent,* and the second the *consequent.*

Terms. Antecedent. Consequent.

In comparing numbers with respect to their difference, the question is, *how much* is one greater than the other? Their difference affords the true answer, and is the measure of their proportion.

Comparison by difference.

In comparing numbers with respect to their quotient, the question is, *how many times* is one greater or less than the other? Their quotient or ratio, is the true answer, and is the measure

Comparison by quotient.

* The term ratio, as now generally used, means the quotient arising from dividing one number by another. We shall use it only in this sense.

Example by difference. By quotient. "Ten times." Examples of this use of the term. Convenient language.

of their proportion. Ten, for example, is 9 greater than 1, if we compare the numbers one and ten by their difference. But if we compare them by their quotient, ten is said to be ten times as great—the language "ten times" having reference to the quotient, which is always taken as the measure of the relative value of two numbers so compared. Thus, when we say, that, the units of our common system of numbers increase in a tenfold ratio, we mean that they so increase that each succeeding unit shall contain the preceding one ten times. This is a convenient language to express a particular relation of two numbers, and is perfectly correct, when used in conformity to an accurate definition.

In what all authors agree: In what disagree. Different methods. Standard the divisor.

§ 165. All authors agree, that the measure of the geometrical proportion, between two numbers, is their ratio; but they are by no means unanimous, nor does each always agree with himself, in the manner of determining this ratio. Some determine it, by dividing the first term by the second; others, by dividing the second term by the first.* All agree, that the *standard*, whatever it may be, should be made the divisor.

* The Encyclopedia Metropolitana, a work distinguished by the excellence of its scientific articles, adopts the latter method.

This leads us to inquire, whether the mind fixes most readily on the first or second number as a standard; that is, whether its tendency is to regard the second number as arising from the first, or the first as arising from the second. What is the best form.

§ 166. All our ideas of numbers begin at one.* This is the starting-point. We conceive of a number only by measuring it with one, as a standard. One is primarily in the mind before we acquire an idea of any other number. Hence, then, the comparison begins at one, which is the standard or unit, and all other numbers are measured by it. When, therefore, we inquire what is the relation of one to any other number, as eight, the idea presented is, how many times does eight contain the standard?

Origin of numbers. How we conceive of a number. Where the comparison begins. The idea presented.

We measure by this standard, and the ratio is the result of the measurement. In this view of the case, the standard should be the first number named, and the ratio, the quotient of the second number divided by the first. Thus, the ratio of 2 to 6 would be expressed by 3, three being the number of times which 6 contains 2.

Standard. Ratio. What they should be. Example.

---

* Section 104.

Other reasons for this method of comparison.

§ 167. The reason for adopting this method of comparison will appear still stronger, if we take fractional numbers. Thus, if we seek the relation between one and one-half, the mind immediately looks to the *part* which one-half is of one, and this is determined by dividing one-half by 1; that is, by dividing the second by the first: whereas, if we adopt the other method, we divide our standard, and find a quotient 2.

Comparison of unity with fractions.

Geometrical proportion.

§ 168. It may be proper here to observe, that while the term "geometrical proportion" is used to express the relation of two numbers, compared by their ratio, the term, "A geometrical proportion," is applied to four numbers, in which the ratio of the first to the second is the same as that of the third to the fourth. Thus,

A geometrical proportion defined.

Example.

$$2 : 4 :: 6 : 12,$$

is a geometrical proportion, of which the ratio is 2.

Further advantages.

§ 169. We will now state some further advantages which result from regarding the ratio as the quotient of the second term divided by the first.

Questions in the Rule of Three:

Every question in the Rule of Three is a geometrical proportion, excepting only, that the last term is wanting. When that term is found, the geometrical proportion becomes complete. In all such proportions, the first term is used as the divisor. Further, for every question in the Rule of Three, we have this clear and simple solution: viz. that, the unknown term or answer, is equal to the third term multiplied by the ratio of the first two. This simple rule, for finding the fourth term, cannot be given, unless we define ratio to be the quotient of the second term divided by the first. Convenience, therefore, as well as general analogy, indicates this as the proper definition of the term ratio.

Their nature.

How solved.

This rule depends on the definition of Ratio.

§ 170. Again, all authors, so far as I have consulted them, are uniform in their definition of the ratio of a geometrical progression: viz. that it is the quotient which arises from dividing the second term by the first, or any other term by the preceding one. For example, in the progression

This definition of ratio is used by all authors, in one case:

2 : 4 : 8 : 16 : 32 : 64, &c.,

Example:

all concur that the ratio is 2; that is, that it is the quotient which arises from dividing the second term by the first: or any other term by the preceding term. But a geometrical progression differs from a geometrical proportion only in

in which they all agree.

The same should take place in every proportion; for they are all the same.

this: in the former, the ratio of any two terms is the same; while in the latter, the ratio of the first and second is different from that of the second and third. There is, therefore, no essential difference in the two proportions.

Why, then, should we say that in the proportion

$$2 : 4 :: 6 : 12,$$

Examples.

the ratio is the quotient of the first term divided by the second; while in the progression

$$2 : 4 : 8 : 16 : 32 : 64, \&c.,$$

the ratio is defined to be the quotient of the second term divided by the first, or of any term divided by the preceding term?

Wherein authors have departed from their definitions:

As far as I have examined, all the authors who have defined the ratio of two numbers to be the quotient of the first divided by the second, have departed from that definition in the case of a geometrical progression. They have there used the word ratio, to express the quotient of the second term divided by the first, and this without any explanation of a change in the definition.

How used ratio.

Other instances in which the definition of

Most of them have also departed from their definition, in informing us that "numbers increase from right to left in a tenfold ratio," in which the term ratio is used to denote the quotient of the second number divided by the first. The definition of ratio is thus departed from, and the idea of it becomes confused. Such discrepancies cannot but introduce confusion into the minds of learners. The same term should always be used in the same sense, and have but a single signification. Science does not permit the slightest departure from this rule. I have, therefore, adopted but a single signification of ratio, and have chosen that one to which all authors, so far as I know, have given their sanction; although some, it is true, have also used it in a different sense.

Ratio is not adhered to.

Consequences.

What science demands.

The definition adopted.

§ 171. One important remark on the subject of proportion is yet to be made. It is this:

Important Remark.

*Any two numbers which are compared together, either by their difference or quotient, must be of the same kind: that is, they must either have the same unit, as a basis, or be susceptible of reduction to the same unit.*

Numbers compared must be of the same kind.

For example, we can compare 2 pounds with 6 pounds: their difference is 4 pounds, and their ratio is the abstract number 3. We can also compare 2 feet with 8 yards: for, although the unit 1 foot is different from the unit 1 yard, still 8 yards are equal to 24 feet. Hence, the differ-

Examples relating to Arithmetical and Geometrical Proportion.

ence of the numbers is 22 feet, and their ratio the abstract number 12.

Numbers with different units cannot be compared.

On the other hand, we cannot compare 2 dollars with 2 yards of cloth, for they are quantities of different kinds, not being susceptible of reduction to a common unit.

Abstract numbers may be compared.

Simple or abstract numbers may always be compared, since they have a common unit 1.

---

## SECTION IV.

### APPLICATIONS OF THE SCIENCE OF ARITHMETIC.

Arithmetic: In what a science.

§ 172. ARITHMETIC is both a science and an art. It is a science in all that relates to the properties, laws, and proportions of numbers.

Science defined.

The science is a collection of those connected processes which develop and make known the laws that regulate and govern all the operations performed on numbers.

What the science performs.

§ 173. Arithmetic is an art, in this: the science lays open the properties and laws of numbers, and furnishes certain principles from which practical and useful rules are formed, applicable in the mechanic arts and in business transactions. The art of Arithmetic consists in the judicious and skilful application of the principles of the science; and the rules contain the directions for such application.

In what the art consists.

§ 174. In explaining the science of Arithmetic, great care should be taken that the analysis of every question, and the reasoning by which the principles are proved, be made according to the strictest rules of mathematical logic.

In explaining the science: what necessary.

Every principle should be laid down and explained, not only with reference to its subsequent use and application in arithmetic, but also, *with reference to its connection with the entire mathematical science*—of which, arithmetic is the elementary branch.

How each principle should be stated.

§ 175. That analysis of questions, therefore, where cost is compared with quantity, or quantity with cost, and which leads the mind of the learner to suppose that a ratio exists between quantities that have not a common unit, is, without explanation, certainly faulty as a process of science.

What questions are faulty.

For example: if two yards of cloth cost 4 dollars, what will 6 yards cost at the same rate?

Example.

Analysis: *Analysis.*—Two yards of cloth will cost twice as much as 1 yard: therefore, if two yards of cloth cost 4 dollars, 1 yard will cost 2 dollars. Again: if 1 yard of cloth cost 2 dollars, 6 yards, being six times as much, will cost six times two dollars, or 12 dollars.

Satisfactory to a child. Now, this analysis is perfectly satisfactory to a child. He perceives a certain relation between 2 yards and 4 dollars, and between 6 yards and 12 dollars: indeed, in his mind, he *compares* these numbers together, and is perfectly satisfied with the result of the comparison.

Reason why it is defective. Advancing in his mathematical course, however, he soon comes to the subject of proportions, treated as a science. He there finds, greatly to his surprise, that he cannot compare together numbers which have different units; and that his *antecedent* and *consequent* must be of the same kind. He thus learns that the whole system of analysis, based on the above method of comparison, is not in accordance with the principles of science.

True analysis: What, then, is the true analysis? It is this: 6 yards of cloth being 3 times as great as 2 yards, will cost three times as much: but 2 yards cost 4 dollars; hence, 6 yards will cost 3 times 4, or 12 dollars.

More scientific. If this last analysis be not as simple as the first, it is certainly more strictly scientific; and when once learned, can be applied through the whole range of mathematical science.

Its advantages.

§ 176. There is yet another view of this question which removes, to a great degree, if not entirely, the objections to the first analysis. It is this:

Reasons in favor of the first analysis.

The proportion between 1 yard of cloth and its cost, two dollars, cannot, it is true, as the units are now expressed, be measured by a ratio, according to the mathematical definition of a ratio. Still, however, between 1 and 2, *regarded as abstract numbers*, there is the same relation existing as between the numbers 6 and 12, also *regarded as abstract*. Now, by leaving out of view, for a moment, the units of the numbers, and finding 12 as an abstract number, and then assigning to it its proper unit, we have a correct analysis, as well as a correct result.

Numbers must be regarded as abstract:

The analysis then correct.

§ 177. It should be borne in mind, that practical arithmetic, or arithmetic as an art, selects from all the principles of the science, the materials for the construction of its rules and the proofs of its methods. As a mere branch of practical knowledge, it cares nothing about the forms or methods of investigation—it demands

How the rules of arithmetic are formed.

What practical knowledge demands.

the fruits of them all, in the most concentrated and practical form. Hence, the best rule of art, which is the one most easily applied, and which reaches the result by the shortest process, is not always constructed after those methods which science employs in the development of its principles.

Best rule of art.

Definition of multiplication.

For example, the definition of multiplication is, that it is the process of taking one number, called the multiplicand, as many times as there are units in another called the multiplier. This definition, as one of science, requires two things.

What it demands.

First.

1st. That the multiplier be an abstract number; and,

Second.

2dly. That the product be of the same kind as the multiplicand.

May be differently considered as furnishing a rule of art.

These two principles are certainly correct, and relating to arithmetic as *a science,* are universally true. But are they universally true, in the sense in which they would be understood by learners, when applied to arithmetic as a mixed subject, that is, a science and an art? Such an application would certainly exclude a large class of practical rules, which are used in the applications of arithmetic, without reference to particular units.

Examples of such applications.

For example, if we have feet in length to be multiplied by feet in height, we must exclude the question as one to which arithmetic is not applicable; or else we must multiply, as indeed we do, without reference to the unit, and then assign a proper unit to the product.

When the three factors are lines.

If we have a product arising from the three factors of length, breadth, and thickness, the unit of the first product and the unit of the final product, will not only be different from each other, but both of them will be different from the unit of the given numbers. The unit of the given numbers will be a unit of length, the unit of the first product will be a square, and that of the final product, a cube.

The different units.

Other examples.

§ 178. Again, if we wish to find, by the best practical rule, the cost of 467 feet of boards at 30 cents per foot, we should multiply 467 by 30, and declare the cost to be 14010 cents, or $140,10.

Considered as a question of science.

Now, as a question of science, if you ask, can we multiply feet by cents? we answer, certainly not. If you again ask, is the result obtained right? we answer, yes. If you ask for the analysys, we give you the following:

1 foot of boards : 467 feet : : 30 cents : Answer.

Ratio.

Now, the ratio of 1 foot to 467 feet, is the abstract number 467; and 30 cents being multiplied by this number, gives for the product 14010 cents. But as the product of two numbers is numerically the same, whichever number be used as the multiplier, we know that 467 multiplied by 30, gives the same number of units as 30 multiplied by 467: hence, the first rule for finding the amount is correct.

Product of two numbers.

The first rule correct.

Scientific investigation:

Practica rule:

Their difference: in what it consists.

§ 179. I have given these illustrations to point out the difference between a process of scientific investigation and a practical rule.

The first should always present the ideas of the subject in their natural order and connection, while the other should point out the best way of obtaining a desired result. In the latter, the steps of the process may not conform to the order necessary for the investigation of principles; but the *correctness of the result* must be susceptible of rigorous proof. Much needless and unprofitable discussion has arisen on many of the processes of arithmetic, from confounding a principle of science with a rule of mere application.

Causes of error.

## SECTION V.

### METHODS OF TEACHING ARITHMETIC CONSIDERED.

---

#### ORDER OF THE SUBJECTS.

§ 180. It has been well remarked by Cousin, the great French philosopher, that "As is the method of a philosopher, so will be his system; and the adoption of a method decides the destiny of a philosophy."

Cousin.

Method decides Philosophy.

What is said here of philosophy in general, is eminently true of the philosophy of mathematical science; and there is no branch of it to which the remark applies, with greater force, than to that of arithmetic. It is here, that the first notions of mathematical science are acquired. It is here, that the mind wakes up, as it were, to the consciousness of its reasoning powers. Here, it acquires the first knowledge of the abstract—separates, for the first time, the pure ideal from the actual, and begins to reflect and reason on pure mental conceptions. It is, therefore, of the highest importance that these first thoughts be impressed on the mind in their natural and proper

True in science.

Why important in Arithmetic.

First thoughts should be rightly impressed.

Faculties to be cultivated.

order, so as to strengthen and cultivate, at the same time, the faculties of apprehension, discrimination, and comparison, and also improve the yet higher faculty of logical deduction.

First point: method of presenting the subject.

§ 181. The first point, then, in framing a course of arithmetical instruction, is to determine the method of presenting the subject. Is there any thing in the *nature* of the subject itself, or the connection of its parts, that points out the order in which these parts should be studied? Do the laws of science demand a particular order; or are the parts so loosely connected, as to render it a matter of indifference where we begin and where we end? A review of the analysis of the subject will aid us in this inquiry.

Laws of science: what do they require?

Basis of the science of numbers.

§ 182. We have seen* that the science of numbers is based on the unit 1. Indeed, the whole science consists in developing, explaining, and illustrating the laws by which, and through which, we operate on this unit. There are three classes of operations performed on the unit one.

In what the science consists.

Three classes of operations.

1st. To increase the unit.

1st. To increase it according to certain scales,

* Section 104.

forming the classes of simple and denominate numbers;

2d. To divide it in any way we please, forming the decimal and vulgar fractions; and,

2d. To divide it.

3d. To compare it with all the numbers which come from it; and then those numbers with each other. This embraces proportions, of which the Rule of Three is the principal branch.

3d. To compare it.

There is yet a fourth branch of arithmetic; viz. the application of the principles and of the rules drawn from them, in the mechanic arts and in the ordinary transactions of business. This is called the Art, or practical part, of Arithmetic. (See Arithmetical Diagram facing page 117.)

Fourth branch

Practical applications;

these the art.

Now, if this analysis be correct, it establishes the order in which the subjects of arithmetic should be taught.

Analysis establishes the order.

## INTEGER UNITS.

§ 183. We begin first with the unit 1, and increase it according to the scale of tens, forming the common system of integer numbers. We then perform on these numbers the operations of the five ground rules; viz. numerate them, add them, subtract them, multiply and divide them.

Unit one increased according to the scale of tens.

Operations performed.

Next increase it according to varying scales.

We next increase the unit 1 according to the varying scales of the denominate numbers, and thus produce the system, called Denominate or Concrete Numbers; after which we perform upon this class all the operations of the five ground rules.

What order the law of exact science requires.

Reason for this.

Reason for departing from it.

§ 184. It may be well to observe here, that the law of exact science requires us to treat the denominate numbers first, and the numbers of the common system afterwards; for, the common system is but a variety of the class of denominate numbers; viz. that variety, in which the scale is the scale of tens, and unvarying. But as *some knowledge of a subject must precede all generalization,* we are obliged to begin the subject of arithmetic with the simplest element.

## FRACTIONAL UNITS.

Divisions of the unit.

General method.

Method according to scale of tens.

§ 185. We now pass to the second class of operations on the unit 1; viz. the divisions of it. Here we pursue the most general method, and divide it arbitrarily; that is, into any number of equal parts. We then observe that the division of it, according to the scale of tens, is but a particular case of the general law of division. We then perform, on the fractional units which thus arise, all the operations of the five ground rules.

Operations performed.

RATIO,—OR RULE OF THREE.

§ 186. Having considered the two subjects of integer and fractional units, we come next to the comparison of numbers with each other.

Subjects considered.

This branch of arithmetic develops all the relative properties of numbers, resulting from their inequality.

What this branch develops.

The method of arrangement, indicated above, presents all the operations of arithmetic in connection with the unit 1, which certainly forms the basis of the arithmetical science.

What the arrangement does.

Besides, this arrangement draws a broad line between the science of arithmetic and its applications; a distinction which it is very important to make. The separation of the principles of a science from their applications, so that the learner shall clearly perceive what is theory and what practice, is of the highest importance. Teaching things separately, teaching them well, and pointing out their connections, are the golden rules of all successful instruction.

What it does further.

Theory and practice should be separated.

Golden rules for teaching.

§ 187. I had supposed, that the place of the

Rule of Three, among the branches of arithmetic, had been fixed long since. But several authors of late, have placed most of the practical subjects *before* this rule—giving precedence, for example, to the subjects of Percentage, Interest, Discount, Insurance, &c. It is not easy to discover the motive of this change. It is certain that the proportion and ratio of numbers are parts of the *science* of arithmetic; and the properties of numbers which they unfold, are indispensably necessary to a clear apprehension of the principles from which the practical rules are constructed.

Differences in arrangement; In what they consist. Ratio part of the science. Should precede applications.

We may, it is true, explain each example in Percentage, Interest, Discount, Insurance, &c., by a separate analysis. But this is a matter of much labor; and besides, does not conduct the mind to any general principle, on which all the operations depend. Whereas, if the Rule of Three be explained, before entering on the practical subjects, it is a great aid and a powerful auxiliary in explaining and establishing all the practical rules. If the Rule of Three is to be learned at all, should it not rather precede than follow its applications? It is a great point, in instruction, to lay down a general principle, as early as possible, and then connect with it, and with each other, all the subor-

Cannot well change the order. Advantages of first explaining the Rule of Three. The great principle of instruction.

dinate principles, with their applications, which flow from it.

### PRACTICAL PART.

§ 188. We come next to the 4th division; viz. the applications of arithmetic. Applications of arithmetic.

Under the classification which we have indicated, all the principles of the science will have been mastered, when the pupil reaches this stage of his progress. His business will now be with the application of principles, and no longer in the study and development of the principles themselves. The unity and simplicity of this method of classification, may be made more apparent, by the aid of the arithmetical diagram which faces page 117. What has been done. What remains to be done. Unity of the classification.

May we not then conclude that the subjects of arithmetic should be presented in the following order: How the subjects should be presented.

1st. All the methods of treating integer numbers, whether formed from the unit 1 according to the scale of tens, or according to varying scales; 1st. Integer numbers.

2d. All the methods of treating fractional unities, whether derived from the unit 1 according to the scale of tens, or according to varying scales; 2d. Fractions.

3d. Rule of Three. 3d. The proportion and ratios of numbers; and,

4th. Applications. 4th. The applications of the science of numbers to practical and useful objects.

OBJECTIONS TO THIS CLASSIFICATION ANSWERED.

Two objections to this classification. § 189. It has been urged that Common or Vulgar Fractions should be placed "immediately after Division, for *two reasons*."

First. "*First*, they arise from division, being in fact *unexecuted* division."

Second. "*Second*, in Reduction and the Compound Rules, it is often necessary to multiply and divide fractions, to add and subtract them, also to carry for them, unless perchance the examples are constructed for the *occasion*, and with *special reference* to *avoiding* these difficulties."

These are all. These, I believe, are all the objections that have been, or can be urged against the classification which I have suggested. I give them in full, because I wish the subject of arrangement to be fully considered and discussed. It should be our main object to get at the best possible system of classification, and not to waste our efforts in ingenious arguments in the support of a favorite one. We will consider these objections separately.

Given in full. What should be our object. To be considered separately.

It is certainly true, that fractions "arise from division," but it is as certainly not true, that they are "*unexecuted* divisions;" and this last idea has involved the subject in much perplexity and difficulty.

Fractions arise from division.

The most elementary idea of a fraction, arises from the division of a single thing into two equal parts, each of which is called a half. Now, we get no idea of this half unless we consider the division *perfected.* And indeed, the method of teaching shows this. For, we cannot impress the idea of a half on the mind of a child, until we have actually divided in his presence the apple (or something else regarded as a unit), and exhibited the parts separately to his senses; and all other fractions must be learned by a like reference to the unit 1. Hence, we can form no notion of a fraction, except on the supposition of a *perfected division.*

The elementary idea is obtained by perfecting the division:

Example;

And not otherwise.

If the term, "*unexecuted* division," applies to the numerator of the expression, and not to the unit of the fraction, the idea is still more involved. For, nothing is plainer than that we can form no distinct *notion of a result,* so long as the process on which it depends cannot be executed. The vague impression that there is something hanging about a fraction that cannot be *quite* reached, has involved the subject in a

"Unexecuted division" does not apply to the numerator.

That a fraction cannot be *quite reached*, has

occasioned difficulty.

mysterious terror; and the boy approaches it with the same feeling which a mariner does a rocky and dangerous coast, of which he has neither map nor chart to guide him. But present to the mind of the pupil the distinct idea, that a fraction is one or more equal parts of unity, and that every such part is a *perfect whole, having a certain relation to the thing from which it was derived*, and all the mist is cleared away, and his mind divides the unit into any number of equal parts, with the same facility as the knife divides the apple.

Every fraction has a fixed relation to unity.

The *form of expression* for a fraction, and for an *unexecuted* division, is indeed the same, but the interpretation of this expression, as used for one or the other, is entirely different. In our common language, the same word is not always the sign of the same idea; and in science, the same symbol often expresses very different things.

Form the same as for an unexecuted division.

A sign may express different things.

For example, $\frac{3}{7}$, as an expression in fractions, means, that something regarded as a whole has been divided in 7 equal parts, and that 3 of those parts are taken. As a result of division, it means that the integer number 3 is to be divided into 7 equal parts. Now, it cannot be assumed, as a self-evident fact, that three of the parts of the first division are equal to 1 part of the second;

Example illustrating h ese principles.

What cannot be assumed.

and if this fact be made the basis of a system of fractions, the mind of a child will go through that system in the dark. *The basis of every system should be an elementary idea.*

The basis of every system should be an elementary idea.

§ 190. The second objection, as far as it goes, is valid. In all the tables of denominate numbers, fractions occur five times; viz. twice in Long Measure, where $5\frac{1}{2}$ yards make 1 rod, and $69\frac{1}{2}$ statute miles 1 degree; once in Cloth Measure, where $2\frac{1}{4}$ inches make 1 nail; once in Square Measure, where $30\frac{1}{4}$ square yards make 1 square rod; and once in Wine Measure, where $31\frac{1}{2}$ gallons make 1 barrel. Now, it were a little better, if these tables had been constructed with integer units. But it should be borne in mind, that the first notions of fractions are given either by oral instruction, or learned from elementary arithmetics. Most of the leading arithmetics are, I believe, preceded by smaller works. These are designed to impart elementary ideas of numbers, so as not to *embarrass the classification of subjects when the scholar is able to enter on a system.* Now, the most elementary of these works conducts the pupil, in fractions, far beyond the point necessary to understand and manage all the fractions which appear in the tables of denominate numbers; and hence, there

Second objection valid;

But of no great weight.

Reasons.

Design of smaller works.

Fractions are partially taught in the elementary works;

May then be used.

is no reason, on that account, to depart from a classification otherwise desirable.

### OBJECTIONS TO THE NEW METHOD.

Objections to the new method considered.

§ 191. Having examined the objections that have been urged against that system of classification of the subjects of arithmetic, which has appeared to me most in accordance with the principles of science, I shall now point out some of the difficulties to be met with in the adoption of the method proposed as a substitute.

First objection.

1st. That method separates the simple and denominate numbers, which, in their general formation, differ from each other only in the scale by which we pass from one unit of value to another.

Second objection.

2d. By thus separating these numbers, it becomes more difficult to point out their connection and teach the important fact, that in all their general properties, and in all the operations to be performed upon them, they differ from each other in no important particular.

Third objection; limitation of the rules.

3d. By placing the denominate numbers after Vulgar Fractions, all the principles and rules in Fractions are limited in their application to *a single class of fractions;* viz. to those fractions *which have the same unit.*

For example, the common rule for addition of fractions, under this classification, is, in substance, the following: "*Reduce the fractions to a common denominator; add their numerators, and place the sum over the common denominator.*"

Examples; showing this.

Rule; not general.

As the subject of denominate numbers has not yet been reached, no allusion can be made to fractions having *different units*. If the learner should happen to understand the rule literally, he would conclude that, the sum of *all fractions* having a common denominator is found by simply adding their numerators and placing the sum over the common denominator. But this cannot, of course, be so, since $\frac{4}{9}$ of a £ and $\frac{5}{9}$ of a shilling make neither one pound nor one shilling.

Have not yet considered fractions having different units;

The rules therefore apply to one class of fractions only.

What appears to me most objectionable in this method, is this: it fails to present the important fact, that no *two fractions* can be blended into one, either by addition or subtraction, unless they are parts of the *same unit,* as wel as *like parts.*

Greatest objection.

By this method of classification most of the difficult questions which arise in fractions are avoided, or else the subject must be resumed after denominate numbers, and that class of questions treated separately.

This method of classification avoids the difficult questions.

What they are.

The class of questions to which I refer, embraces examples like the following:

Add $\frac{5}{7}$ of a day, $\frac{4}{13}$ of an hour, and $\frac{3}{8}$ of a second together.

The subject easily disposed of, but not learnt.

It is certainly true that a boy will make marvellous progress in the text-book, if you limit him to those examples in which the fractions have a common unit. But, will he ever understand the science of fractions unless his mind be steadily and always turned to the unit of the fraction, as the basis? Will he understand the value of one equal part, so as to compare and unite it with another equal part, unless he first apprehends, clearly, the units from which those parts were derived?

Last objection stated.

Difficulty of tracing the connection of the fractions.

4th. By placing the Denominate Numbers between Vulgar and Decimal Fractions, the general subject of fractional arithmetic is broken into fragments. This arrangement makes it difficult to realize that these two systems of numbers differ from each other in no essential particular; that they are both formed from the unit one by the same process, with only a slight modification of the scale of division.

## ARITHMETICAL LANGUAGE.

§ 192. We have seen that the arithmetical alphabet contains ten characters.* From these elements the entire language is formed; and we now propose to show in how simple a manner.

*Arithmetical alphabet.*

The names of the ten characters are the first ten words of the language. If the unit 1 be added to each of the numbers from 1 to 10 inclusive, we find the first ten combinations in arithmetic.† If 2 be added, in like manner, we have the second ten combinations; adding 3, gives us the third ten combinations; and so on, until we have reached one hundred combinations (page 123).

*Names of the characters.*

*First ten combinations.*

*Second ten, and so on for others.*

Now, as we progressed, each set of combinations introduced one additional word, and the results of all the combinations are expressed by the words from two to twenty inclusive.

*Each set giving one additional word.*

§ 193. These one hundred elementary combinations, are all that need be committed to memory; for, every other is deduced from them. They are, in fact, but different spellings of the first nineteen words which follow one. If we extend the words to one hundred, and recollect that

*All that need be committed to memory.*

---

* Section 114. † Section 116.

at one hundred, we begin to repeat the numbers, we see that we have but one hundred words to be remembered for addition; and of these, *all above ten are derivative.* To this number, must of course be added the few words which express the sums of the hundreds, thousands, &c.

Words to be remembered for addition.

Only ten words primitive.

Subtraction:

§ 194. In Subtraction, we also find one hundred elementary combinations; the results of which are to be read.* These results, and all the numbers employed in obtaining them, are expressed by twenty words.

Number of words.

Multiplication:

§ 195. In Multiplication (the table being carried to twelve), we have one hundred and forty-four elementary combinations,† and fifty-nine separate words (already known) to express the results of these combinations.

Number of words.

Division:

§ 196. In Division, also, we have one hundred and forty-four elementary combinations,‡ but use only twelve words to express their results.

Number of words.

Four hundred and eighty-eight elementary combinations.

§ 197. Thus, we have four hundred and eighty-eight elementary combinations. The results of these combinations are expressed by one hundred words; viz. nineteen in addition, ten in subtraction, fifty-nine in multiplication, and twelve

Words used: 19 in addition, 10 in subtraction, 59 in multiplication,

* Section 120. † Section 122. ‡ Section 123.

in division. Of the nineteen words which are employed to express the results of the combinations in addition, eight are again used to express similar results in subtraction. Of the fifty-nine which express the results of the combinations in multiplication, sixteen had been used to express similar results in addition, and one in subtraction; and the entire twelve, which express the results of the combinations in division, had been used to express results of previous combinations. Hence, the results of all the elementary combinations, in the four ground rules, are expressed by sixty-three different words; and they are the only words employed to translate these results from the arithmetical into our common language.

12 in division.

Sixty-three different words in all.

The language for fractional units *is similar in every particular.* By means of a language thus formed we deduce every principle in the science of numbers.

Language the same for fractions.

§ 198. Expressing these ideas and their combinations by figures, gives rise to the language of arithmetic. By the aid of this language we not only unfold the principles of the science, but are enabled to apply these principles to every question of a practical nature, involving the use of figures.

Language of arithmetic:

Its value and use.

But few combinations which change the signification of the figures.

§ 199. There is but one further idea to be presented: it is this,—that there are very few combinations made among the figures, which change, at all, their signification.

Examples.

Selecting any two of the figures, as 3 and 5, for example, we see at once that there are but three ways of writing them, that will at all change their signification.

First:

First, write them by the side of each other - - - - - - - - - - - } 3 5, 5 3.

Second:

Second, write them, the one over the other - - - - - - - - - - } $\frac{3}{5}$, $\frac{5}{3}$.

Third.

Third, place a decimal point before each - - - - - - - - - - - - - } .3, .5.

Learn the language by use.

Now, each manner of writing gives a different signification to both the figures. Use, however, has established that signification, and we know it, as soon as we have learned the language.

Its grammar: Alphabet—words, and their uses.

We have thus explained what we mean by the arithmetical language. Its grammar embraces the names of its elementary signs, or Alphabet,—the formation and number of its words,—and the laws by which figures are connected for the purpose of expressing ideas. We feel that there is simplicity and beauty in this system, and hope it may be useful.

### NECESSITY OF EXACT DEFINITIONS AND TERMS.

§ 200. The principles of every science are a collection of mental processes, having established connections with each other. In every branch of mathematics, the Definitions and Terms give form to, and are the signs of, certain elementary ideas, which are the basis of the science. Between any term and the idea which it is employed to express, the connection should be so intimate, that the one will always suggest the other.

Principles of science.

Definitions and terms:

These definitions and terms, when their significations are once fixed, must always be used in the same sense. The necessity of this is most urgent. For, "*in the whole range of arithmetical science there is no logical test of truth, but in a conformity of the reasoning to the definitions and terms, or to such principles as have been established from them.*"

When once fixed must always be used in the same sense.

Reason.

§ 201. With these principles, as guides, we propose to examine some of the definitions and terms which have, heretofore, formed the basis of the arithmetical science. We shall not confine our quotations to a single author, and shall make only those which fairly exhibit the general use of the terms.

Definitions and terms examined.

It is said,

Number defined. "*Number* signifies a *unit*, or a *collection of units.*"

How expressed. "The common method of expressing numbers is by the *Arabic Notation.* The Arabic method employs the following *ten characters*, or *figures*," &c.

Names of the characters. "The first nine are called *significant* figures, because each one always has a value, or denotes some number."

And a little further on we have,

Figures have values. "The different values which figures have, are called *simple* and *local* values."

Number rightly defined: The definition of Number is clear and correct. It is a general term, comprehending all the phrases which are used, to express, either separately or in connection, one or more things of the same kind. So, likewise, the definition of figures, that they are *characters*, is also right. Also figures.

Definition departed from. But mark how soon these definitions are departed from. The reason given why nine of the figures are called *significant* is, because "each one always has a value, or denotes some number." This brings us directly to the question, whether a figure has a *value;* or, whether it is *a mere representative of value.* Is it a *number* or a *character* to represent number? Is it a *quantity* or *symbol?* It is defined to be *a char-* Has a figure value? It is merely a character:

*acter* which *stands for,* or expresses a number. Has it any other signification? How then can we say that it has a *value*—and how is it possible that it can have a *simple* and a *local value?* The *things* which the figures stand for, may change their value, but not the *figures themselves.* Indeed, it is very difficult for John to perceive how the figure 2, standing in the second place, is ten times as great as the same figure 2 standing in the first place on the right! although he will readily understand, when the arithmetical language is explained to him, that the UNIT of one of these places is ten times as great as that of the other.

Has no value of itself;

but stands for value.

Unit of place.

§ 202. Let us now examine the leading definition or principle which forms the basis of the arithmetical language. It is in these words:

Leading definition.

"*Numbers increase from right to left in a tenfold ratio; that is, each removal of a figure one place towards the left, increases its value ten times.*"

Of number.

Now, it must be remembered, that number has been defined as signifying "a unit, or a collection of units." How, then, can it have a *right hand,* or a *left?* and how can it *increase from right to left in a tenfold ratio?*" The explanation given is, that "*each removal of a*

Does not agree with the definition before given.

Explanation. *figure one place towards the left, increases its value ten times."*

Number, signifying a collection of units, must necessarily increase according to the law by which these units are combined; and that law of increase, whatever it may be, has not the slightest connection with the *figures* which are used to express the numbers.

Increase of numbers has no connection with figures.

Ratio.

Besides, is the term *ratio* (yet undefined), one which expresses an elementary idea? And is the term, a "*tenfold ratio,*" one of sufficient simplicity for the basis of a system?

"Tenfold ratio:"

Does, then, this definition, which in substance is used by most authors, involve and carry to the mind of the young learner, the four leading ideas which form the basis of the arithmetical notation? viz.:

Four leading notions of numbers.

First.

1st. That numbers are expressions for one or more things of the same kind.

Second.

2d. That numbers are expressed by certain characters called figures; and of which there are ten.

Third.

3d. That each figure always expresses as many units as its name imports, and no more.

Fourth.

4th. That the *kind* of thing which a figure expresses depends on the place which the figure occupies, or on the value of the units, indicated in some other way.

PLACE is merely *one* of the forms of language by which we designate the unit of a number, expressed by a figure. The definition attributes this property of place both to number and figures, while it belongs to neither. Place; Its office.

§ 203. Having considered the definitions and terms in the first division of Arithmetic, viz. in notation and numeration, we will now pass to the second, viz. Addition. Definitions in Addition:

The following are the definitions of Addition, taken from three standard works before me:

"The putting together of two or more numbers (as in the foregoing examples), so as to make one *whole number*, is called *Addition*, and the whole number is called the *sum*, or *amount*." First.

"ADDITION is the collecting of numbers together to find their sum." Second.

"*The process of uniting two or more numbers together, so as to form one single number, is called* ADDITION." Third.

"The *answer*, or the number thus found, is called the *sum*, or *amount*."

Now, is there in either of these definitions any test, or means of determining when the pupil gets the thing he seeks for, viz. "the sum of two or more numbers?" No previous definition has been given, in either work, of the Defects. Reason.

term SUM. How is the learner to know what he is seeking for, unless that thing be defined?

No principle as a standard. Suppose that John be required to find the sum of the numbers 3 and 5, and pronounces it to be 10. How will you correct him, by showing that he has not conformed to the definitions and rules? You certainly cannot, because *you have established no test of a correct process.*

But, if you have previously defined SUM to be a number which *contains as many units* as there are in all the numbers added: or, if you say,

Correct definition; "Addition is the process of uniting two or more numbers, *in such a way*, that all the units which they contain may be expressed by a single number, called the sum, or sum total;" you will then have a *test* for the correctness of the process of Addition; viz. Does the number, which you call the sum, contain as many units as there are in all the numbers added? The answer to this question will show that John is wrong.

Gives a test.

Definitions of fractions. § 204. I will now quote the definitions of Fractions from the same authors, and in the same order of reference.

First. "We have seen, that numbers expressing *whole* things, are called *integers*, or *whole* numbers; but that, in division, it is often necessary to *divide* or *break* a whole thing into *parts*, and that these parts are called *fractions*, or *broken* numbers."

"Fractions are parts of an integer." Second.

"*When a number or thing is divided into equal parts, these parts are called* FRACTIONS." Third.

Now, will either of these definitions convey to the mind of a learner, a distinct and exact idea of a fraction?

The term "fraction," as used in Arithmetic, means one or more equal parts of something regarded as a whole: *the parts to be expressed in terms of the thing divided* CONSIDERED AS A UNIT. There are three prominent ideas which the mind must embrace: Term fraction defined. Ideas expressed:

1st. That the thing divided be regarded as a standard, or unity; First.

2d. That it be divided into equal parts; Second.

3d. That the parts be expressed in terms of the thing divided, regarded as a unit. Third.

These ideas are referred to in the latter part of the first definition. Indeed, the definition would suggest them to any one acquainted with the subject, but not, we think, to a learner. The definitions examined:

In the second definition, neither of them is hinted at. Take, for example, the integer number 12, and no one would say that any one part of this number, as 2, 4, or 6, is a fraction. Is a fraction part of an integer?

Third definition; The third definition would be perfectly accurate, by inserting after the word "thing," the words, "regarded as a whole." It very clearly expresses the idea of equal parts, but does not present the idea strongly enough, that the thing divided must be regarded as unity, and that the parts must be expressed in terms of this unity.

In what defective.

§ 205. I have thus given a few examples, illustrating the necessity of accurate definitions and terms. Nothing further need be added, except the remark, that terms should always be used in the same sense, precisely, in which they are defined.

Necessity of exact terms.

Objection to exactness of thought and language. To some, perhaps, these distinctions may appear over-nice, and matters of little moment. It may be supposed that a general impression, imparted by a language reasonably accurate, will suffice very well; and that it is hardly worth while to pause and weigh words on a nicely-adjusted balance.

Any such notions, permit me to say, will lead to fatal errors in education.

Definitions in mathematics. It is in mathematical science alone that words are the signs of exact and clearly-defined ideas. It is here only that we can see, as it were, the very thoughts through the transparent words by which they are expressed. If the words of the definitions are not such as convey to the mind of the learner, the fundamental ideas of the science, he cannot reason upon these ideas; for, he does not apprehend them; and the great reasoning faculty, by which all the subsequent principles of mathematics are developed, is entirely unexercised.*

Must be exact to reason correctly.

It is not possible to cultivate the habit of accurate thinking, without the aid and use of exact language. No mental habit is more useful than that of tracing out the connection between ideas and language. In Arithmetic, that connection can be made strikingly apparent. Clear, distinct ideas—diamond thoughts—may be strung through the mind on the thread of science, and each have its word or phrase by which it can be transferred to the minds of others.

Cannot otherwise cultivate habits of thought.

Connection between words and thoughts in arithmetic.

HOW SHOULD THE SUBJECTS BE PRESENTED?

§ 206. Having considered the natural connection of the subjects of arithmetic with each other, as branches of a single science, based on a single unit; and having also explained the necessity of a perspicuous and accurate lan-

What has been considered.

* Section 200.

How ought the subjects to be presented. guage; we come now to that important inquiry, How ought those subjects to be presented to the mind of a learner? Before answering this question, we should reflect, that two important objects should be sought after in the study of arithmetic:

Two objects in studying arithmetic:

First. 1st. To train the mind to habits of clear, quick, and accurate thought—to teach it to apprehend distinctly—to discriminate closely—to judge truly—and to reason correctly; and,

Second. 2d. To give, in abundance, that *practical* knowledge of the use of figures, in their various applications, which shall illustrate the striking fact, that *the art of arithmetic is the most important art of civilized life—being, in fact, the foundation of nearly all the others.*

Art of arithmetic.

How first impressions are made. § 207. It is certainly true, that most, if not all the elementary notions, whether abstract or practical—that is, whether they relate to the science or to the art of arithmetic, must be made on the mind by means of sensible objects. Because of this fact, many have supposed that the *processes* of *reasoning* are all to be conducted by the same sensible objects; and that every abstract principle of science is to be developed and established by means of sofas, chairs, apples, and horses. There seems to be

Is reasoning to be conducted by sensible objects?

an impression that because blocks are useful aids in teaching the alphabet, that, *therefore* they can be used advantageously in reading Milton and Shakspeare. This error is akin to that of attempting to teach practically, Geography and Surveying in connection with Geometry, by calling the angles of a rectangle, north, south, east, and west, instead of simply designating them by the letters A, B, C, and D.

Sensible objects useful in acquiring the simplest elements:

Error of carrying them beyond.

This false idea, that every principle of science must be *learned practically,* instead of being *rendered practical by its applications,* has been highly detrimental both to science and art.

False idea:

Its effects.

A mechanic, for example, knowing the height of his roof and the width of his building, wishes to cut his rafters to the proper length. If he calls to his aid the established, though *abstract principles* of science, he finds the length of his rafter, by the well-known relation between the hypothenuse and the two sides of a right-angled triangle. If, however, he will learn nothing except *practically,* he must raise his rafter to the roof, measure it, and if it be too long cut it off, if too short, splice it. This is the practical way of *learning* things.

Example of the application of an abstract principle:

Of learning practically.

The truly practical way, is that in which skill is guided by science.

True practical.

Do the principles above stated find any appli-

Can be applied.

cation in considering the question, How should arithmetic be taught? Certainly they do. If arithmetic be both a science and an art, it should be so taught and so learned.

Principles of science: What they are: Wise to use them.

§ 208. The principles of every science are general and abstract truths. They are mere ideas, primarily acquired through the senses by experience, and generalized by processes of reflection and reasoning; and when understood, are certain guides in every case to which they are applicable. If we choose to do without them, we may. But is it wise to turn our heads from the guide-boards and explore every road that opens before us?

When and how they should be taught. The methods pointed out. Principles to be impressed.

Now, in the study of arithmetic those principles of science, applicable to classes of cases, should always be taught at the earliest possible moment. The mind should never be forced through a long series of examples, without explanation. One or two examples should always precede the statement of an abstract principle, or the laying down of a rule, so as to make the language of the principle or rule intelligible. But to carry the learner forward through a series of them, before the principle on which they depend has been examined and stated, is forcing the mind to advance mechanically—it is lifting up the rafter to measure it, when its exact length could be easily determined by a rule of science.

As most of the instruction in arithmetic must be given with the aid of books, we feel unable to do justice to this branch of the subject without submitting a few observations on the nature of text-books and the objects which they are intended to answer.

Books: Necessity for treating of them.

## TEXT-BOOKS.

§ 209. A text-book should be an aid to the teacher in imparting instruction, and to the learner in acquiring knowledge.

Text-book:

It should present the subjects of knowledge in their proper order, with the branches of each subject classified, and the parts rightly arranged. No text-book, on a subject of general knowledge, can contain all that is known of the subject on which it treats; and ordinarily, it can contain but a very small part. Hence, the subjects to be presented, and the extent to which they are to be treated, are matters of nice discrimination and judgment, about which there must always be a diversity of opinion.

What it should be. Selection of subjects necessary. Difficulties of selection.

§ 210. The subjects selected should be leading ones, and those best calculated to unfold, ex-

Subjects:

plain, and illustrate the principles of the science. How presented. They should be so presented as to lead the mind to analyze, discriminate, and classify; to see each principle separately, each in its combination with others, and all, as forming an harmonious whole. Too much care cannot be bestowed in forming the *suggestive method of arrangement:* that is, to place the ideas and principles in such a connection, that *each step shall prepare the mind of the learner for the next in order.*

Suggestive method: Reason for.

Object of a text-book:

§ 211. A text-book should be constructed for the purpose of furnishing the learner with the keys of knowledge. It should point out, explain, and illustrate by examples, the methods of investigating and examining subjects, but should leave the mind of the learner free from the restraints of minute detail. To fill a book with the analysis of simple questions, which any child can solve in his own way, is to constrain and force the mind at the very point where it is capable of self-action. To do that for a pupil, which he can do for himself, is most unwise.

Nature; Useless detail;

Should not be historical.

§ 212. A text-book on a subject of science should not be historical. At first, the minds of children are averse to whatever is abstract, because what is abstract demands thought, and thinking is mental labor from which untrained minds turn away. If the thread of science be broken by the presentation of facts, having no connection with the argument, the mind will leave the more rugged path of the reasoning, and employ itself with what requires less effort and labor. Reasons.

The optician, in his delicate experiments, excludes all light except the beam which he uses: so, the skilful teacher excludes all thoughts excepting those which he is most anxious to impress. Illustration.

As a general rule, subject of course to some exceptions, but one method for each process should be given. The minds of learners should not be confused. If several methods are given, it becomes difficult to distinguish the reasonings applicable to each, and it requires much knowledge of a subject to compare different methods with each other. One method. Reasons.

§ 213. It seems to be a settled opinion, both among authors and teachers, that the subject of arithmetic can be best presented by means of three separate works. For the sake of distinction, we will designate them the First, Second, and Third Arithmetics. How the subject is divided.

We will now explain what we suppose to be the proper construction of each book, and the object for which each should be designed.

### FIRST ARITHMETIC.

First Arithmetic: Its importance.

§ 214. This book should give to the mind its first direction in mathematical science, and its first impulse in intellectual development. Hence, it is the most important book of the series. Here, the faculties of apprehension, discrimination, abstraction, classification and comparison, are brought first into activity. Now, to cultivate and develop these faculties rightly, we must, at first, present every new idea by means of a sensible object, and then immediately drop the object and pass to the abstract thought.

How the subjects must be presented.

Order of the ideas.

We must also present the ideas consecutively; that is, in their proper order; and by the mere *method of presentation* awaken the comparative and reasoning faculties. Hence, every lesson should contain a given number of ideas. The ideas of each lesson, beginning with the first, should advance in regular gradation, and the lessons themselves should be regular steps in the progress and development of the arithmetical science.

Construction of the lessons.

First lesson.

§ 215. The first lesson should merely contain representations of sensible objects, placed opposite names of numbers, to give the impression of the meanings of these names: thus,

What it should contain.

| | |
|---|---|
| One - - - - - - - - - - | * |
| Two - - - - - - - - - | * * |
| Three - - - - - - - - | * * * |
| &c. | &c. |

And with young pupils, more striking objects should be substituted for the stars.

In the second lesson, the words should be replaced by the figures: thus,

Second lesson.

| | |
|---|---|
| 1 - - - - - - - - - - - | * |
| 2 - - - - - - - - - - | * * |
| 3 - - - - - - - - - - | * * * |
| &c. | &c. |

In the third lesson, I would combine the ideas of the first two, by placing the words and figures opposite each other: thus,

Third lesson

| | | | |
|---|---|---|---|
| One - - - - | 1 | Four - - - - | 4 |
| Two - - - - | 2 | Five - - - - | 5 |
| Three - - - | 3 | Six - - - - | 6 |
| &c. | &c. | &c. | &c. |

The Roman method of representing numbers should next be taught, making the fourth lesson: viz.,

Fourth lesson.

Roman method.

| | | | |
|---|---|---|---|
| One - - - - | I. | Four - - - | IV. |
| Two - - - - | II. | Five - - - | V. |
| Three - - | III. | Six - - - | VI. |
| &c. | &c. | &c. | &c. |

First ten combinations:

§ 216. We come now to the first ten combinations of numbers, which should be given in a separate lesson. In teaching them, we must, of course, have the aid of sensible objects. We teach them thus:

How taught by things:

| | | | |
|---|---|---|---|
| One | and | one | are how many? |
| * | | * | |
| One | and | two | are how many? |
| * | | * * | |
| One | and | three | are how many? |
| * | | * * * | |
| &c. | | &c. | &c., |

through all the combinations: after which, we pass to the abstract combinations, and ask, one and one are how many? one and two, how many? one and three, &c.; after which we express the results in figures.

How in the abstract.

We would then teach in the same manner, in a separate lesson, the second ten combinations; then the third, fourth, fifth, sixth, seventh, eighth, ninth, and tenth. In the teaching of these combinations, only the words from one to twenty will have been used. We must then teach the

Second ten combinations.

Words used.

combinations of which the results are expressed by the words from twenty to one hundred. Further combinations.

§ 217. Having done this, in the way indicated, the learner sees at a glance, the basis on which the system of common numbers is constructed. He distinguishes readily, the unit one from the unit ten, apprehends clearly how the second is derived from the first, and by comparing them together, comprehends their mutual relation. Results. How they appear.

Having sufficiently impressed on the mind of the learner, the important fact, that numbers are but expressions for one or more things of the same kind, the unit mark may be omitted in the combinations which follow. Unit mark omitted.

§ 218. With the single difference of the omission of the unit mark, the very same method should be used in teaching the one hundred combinations in subtraction, the one hundred and forty-four in multiplication, and the one hundred and forty-four in division. Same method in the other rules.

When the elementary combinations of the four ground rules are thus taught, the learner looks back through a series of regular progression, in which every lesson forms an advancing step, and where all the ideas of each lesson have a Results of the method:

mutual and intimate connection with each other. Will not such a system of teaching train the mind to the habit of regarding each idea separately—of tracing the connection between each new idea and those previously acquired—and of comparing thoughts with each other?—and are not these among the great ends to be attained, by instruction?

Are they desirable?

The power they give.

§ 219. It has seemed to me of great importance to use figures in the very first exercises of arithmetic. Unless this be done, the operations must all be conducted by means of sounds, and the pupil is thus taught to regard sounds as the proper symbols of the arithmetical language. This habit of mind, once firmly fixed, cannot be easily eradicated; and when the figures are learned afterwards, they will not be regarded as the representatives of as many things as their names respectively import, but as the representatives merely of familiar sounds which have been before learned.

Figures should be used early.

Reasons.

Consequences of using words only

This would seem to account for the fact, about which, I believe, there is no difference of opinion; that a course of oral arithmetic, extending over the whole subject, without the aid and use of figures, is but a poor preparation for operations on the slate. It may, it is true,

Oral arithmetic:

sharpen and strengthen the mind, and give it development: but does it give it that language and those habits of thought, which turn it into the pathways of science? The language of a science affords the tools by which the mind pries into its mysteries and digs up its hidden treasures. The language of arithmetic is formed from the ten figures. By the aid of this language we measure the diameter of a spider's web, or the distance to the remotest planet which circles the heavens; by its aid, we calculate the size of a grain of sand and the magnitude of the sun himself: should we then abandon a language so potent, and attempt to teach arithmetic in one which is unknown in the higher departments of the science?

What it may do. What it does not do. Language of arithmetic: Its uses. What it performs. Its value.

§ 220. We next come to the question, how the subject of fractions should be presented in an elementary work.

Fractions:

The simplest idea of a fraction comes from dividing the unit one into two equal parts. To ascertain if this idea is clearly apprehended, put the question, How many halves are there in one? The next question, and it is an important one, is this: How many halves are there in one and one-half? The next, How many halves in two? How many in two and a half? In

Simplest idea. How impressed. Next question.

three? Three and a half? and so on to twelve. Results. You will thus evolve all the halves from the units of the numbers from one to twelve, inclusive. We stop here, because the multiplication table goes no further. These combinations First lesson. should be embraced in the first lesson on fractions. That lesson, therefore, will teach the Its extent. relation between the unit 1 and the halves, and point out how the latter are obtained from the former.

Second lesson. § 221. The second lesson should be the first, reversed. The first question is, how many Grades of questions. whole things are there in two halves? Second, How many whole things in four halves? How many in eight? and so on to twenty-four halves, when we reach the extent of the division Extent of the lesson. table. In this lesson you will have taught the pupil to pass back from the fractions to the unit from which they are derived.

Fundamental principles: § 222. You have thus taught the two fundamental principles of all the operations in fractions: viz.

First. 1st. To deduce the fractional units from integer units; and,

Second. 2dly. To deduce integer units from fractional units.

§ 223. The next lesson should explain the law by which the thirds are derived from the units from 1 to 12 inclusive; and the following lesson the manner of changing the thirds into integer units.

Lessons explaining thirds.

The next two lessons should exhibit the same operations performed on the fourth, the next two on the fifth, and so on to include the twelfth.

Fourths and other fractions.

§ 224. This method of treating the subject of fractions has many advantages:

Advantages of the method.

1st. It points out, most distinctly, the relations between the unit 1 and the fractions which are derived from it.

First.

2d. It points out clearly the methods of passing from the fractional to the integer units.

Second.

3d. It teaches the pupil to handle and combine the fractional units, as entire things.

Third.

4th. It reviews the pupil, thoroughly, through the multiplication and division tables.

Fourth.

5th. It awakens and stimulates the faculties of apprehension, comparison, and classification.

Fifth.

§ 225. Besides the subjects already named, the First Arithmetic should also contain the tables of denominate numbers, and collections of simple examples, to be worked on the slate,

What else the First Arithmetic should contain.

Examples, how taught. under the direction of the teacher. It is not supposed that the mind of the pupil is sufficiently matured at this stage of his progress to understand and work by rules.

What should be taught in the First Arithmetic. § 226. In the First Arithmetic, therefore, the pupil should be taught,

1st. The language of figures;

Second. 2d. The four hundred and eighty-eight elementary combinations, and the words by which they are expressed;

Third. 3d. The main principles of Fractions;

Fourth. 4th. The tables of Denominate Numbers; and,

Fifth. 5th. To perform, upon the slate, the elementary operations in the four ground rules.

### SECOND ARITHMETIC.

Second Arithmetic. § 227. This arithmetic occupies a large space in the school education of the country. Many study it, who study no other. It should, there- What it should be. fore, be complete in itself. It should also be eminently practical; but it cannot be made so either by giving it the name, or by multiplying the examples.

Practical application of principle. § 228. The truly practical cannot be the antecedent, but must be the consequent of science.

Hence, that general arrangement of subjects demanded by science, and already explained, must be rigorously followed. Arrangement of subjects.

But in the treatment of the subjects themselves, we are obliged, on account of the limited information of the learners, to adopt methods of teaching less general than we could desire. Reasons for departures.

§ 229. We must here, again, begin with the unit one, and explain the general formation of the arithmetical language, and must also adhere rigidly to the method of introducing new principles or rules by means of sensible objects. This is most easily and successfully done either by an example or question, so constructed as to show the application of the principle or rule. Such questions or examples being used merely for the purpose of illustration, one or two will answer the purpose much better than twenty: for, if a large number be employed, they are regarded as examples for practice, and are lost sight of as illustrations. Besides, it confuses the mind to drag it through a long series of examples, before explaining the principles by which they are solved. One example, wrought under a principle or rule clearly apprehended, conveys to the mind more practical information, than a dozen wrought out as independent

Basis. Method. How carried out. Few examples. Reasons. One example under a rule.

Principle. Practice. exercises. Let the principle precede the practice, in all cases, as soon as the information acquired will permit. This is the golden rule both of art and morals.

Subjects embraced. § 230. The Second Arithmetic should embrace all the subjects necessary to a full view of the science of numbers; and should contain an abundance of examples to illustrate their practical applications. Reading: The reading of numbers, so much (though not too much) dwelt upon, is an invaluable aid in all practical operations.

Its value in Addition: By its aid, in addition, the eye runs up the columns and collects, in a moment, the sum of all the numbers. Subtraction: In subtraction, it glances at the figures, and the result is immediately suggested. Multiplication: In multiplication, also, the sight of the figures brings to mind the result, and it is reached and expressed by one word instead of five. Division. In short division, likewise, there is a corresponding saving of time by reading the results of the operations instead of spelling them. The method of reading should, therefore, be constantly practised, and none other allowed.

## THIRD ARITHMETIC.

§ 231. We have now reached the place where arithmetic may be taught as a science. The pupil, before entering on the subject as treated here, should be able to perform, at least mechanically, the operations of the five ground rules. Third Arithmetic: Preparation for.

Arithmetic is now to be looked at from an entirely different point of view. The great principles of generalization are now to be explained and applied. View of it.

Primarily, the general language of figures must be taught, and the striking fact must then be explained, that the construction of all integer numbers involves but a single principle, viz. *the law of change in passing from one unit to another*. The basis of all subsequent operations will thus have been laid. What is taught primarily. General law:

§ 232. Taking advantage of this general law which controls the formation of numbers, we bring all the operations of reduction under one single principle, viz. this law of change in the unities. Controls formation of numbers

Passing to addition, we are equally surprised and delighted to find the same principle controlling all its operations, and that integer numbers of all kinds, whether simple or denominate, may be added under a single rule. Its value in Addition.

Advantages of knowing a general law. This view opens to the mind of the pupil a wide field of thought. It is the first illustration of the great advantage which arises from looking into the laws by which numbers are constructed. Subtraction. In subtraction, also, the same principle finds a similar application, and a simple rule containing but a few words is found applicable to all the classes of integer numbers.

In multiplication and division, the same striking results flow from the same cause; and General law of numbers: thus this simple principle, viz. *the law of change in passing from one unit of value to another, is the key to all the operations in the four ground rules*, whether performed on simple or denominate numbers. Thus, all the elementary operations Controls every operation. of arithmetic are linked to a single principle, and that one a *mere principle of arithmetical language*. Who can calculate the labor, intellectual and mechanical, which may be saved by a right application of this luminous principle ?

Design of the higher arithmetic: § 233. It should be the design of a higher arithmetic to expand the mind of the learner over the whole science of numbers; to illustrate the most important applications, and to make manifest the connection between the science and the art.

It will not answer these objects if the methods of treating the subject are the same as in the elementary works, where science has to compromise with a want of intelligence. An elementary is not made a higher arithmetic, by merely transferring its definitions, its principles, and its rules into a larger book, in the same order and connection, and arranging under them an apparently new set of examples, though in fact constructed on precisely the same principles.

Its requisites.

Must have a distinctive character.

§ 234. In the four ground rules, particularly (where, in the elementary works, simple examples must necessarily be given, because here they are used both for illustration and practice), the examples should take a wide range, and be so selected and combined as to show their common dependence on the same principle.

Construction of examples in the four ground rules.

§ 235. It being the leading design of a series of arithmetics to explain and illustrate the science and art of numbers, great care should be taken to treat all the subjects, as far as their different natures will permit, according to the same general methods. In passing from one book to another, every subject which has been fully and satisfactorily treated in the one, should be transferred to the other with the fewest pos-

Design of a series.

Subjects to be transferred when fully treated.

How common subjects may be studied.

sible alterations ; so that a pupil shall not have to learn under a new dress that which he has already fully acquired. They who have studied the elementary work should, in the higher one, either omit the common subjects or pass them over rapidly in review.

Reasons.

The more enlarged and comprehensive views which should be given in the higher work will thus be acquired with the least possible labor, and the connection of the series clearly pointed out.

Additional reason stated.

This use of those subjects, which have been fully treated in the elementary work, is greatly preferable to the method of attempting to teach every thing anew : for there must necessarily be much that is common ; and that which teaches no new principle, or indicates no new method of application, should be precisely the same in the higher work as in that which precedes it.

A contrary method leads to confusion:

§ 236. To vary the examples, in form, without changing in the least the principles on which they are worked, and to arrange a thousand such collections under the same set of rules and subject to the same laws of solution, may give a little more mechanical facility in the use of figures, but will add nothing to the stores of arithmetical knowledge. Besides, it deludes the learner with the hope of advancement, and when he reaches the end of his higher arithmetic, he finds, to his amazement, that he has been conducted by the same guides over the same ground through a winding and devious way, made strange by fantastic drapery: whereas, if what was new had been classed by itself, and what was known clothed in its familiar dress, the subject would have been presented in an entirely different and brighter light.

It misleads the pupil:

It complicates the subject.

### CONCLUDING REMARKS.

We have thus completed a full analysis of the language of figures, and of the construction of numbers.

Conclusion.

We have traced from the unit one, all the numbers of arithmetic, whether integer or fractional, whether simple or denominate. We have developed the laws by which they are derived from this common source, and perceived the connections of each class with all the others.

What has been done.

Laws.

We have examined that concise and beautiful language, by means of which numbers are made available in rendering the results of science practically useful; and we have also considered the best methods of teaching this great subject —the foundation of all mathematical science— and the first among the useful arts.

Analysis of the language.

Methods of teaching indicated.

Importance of the subject.

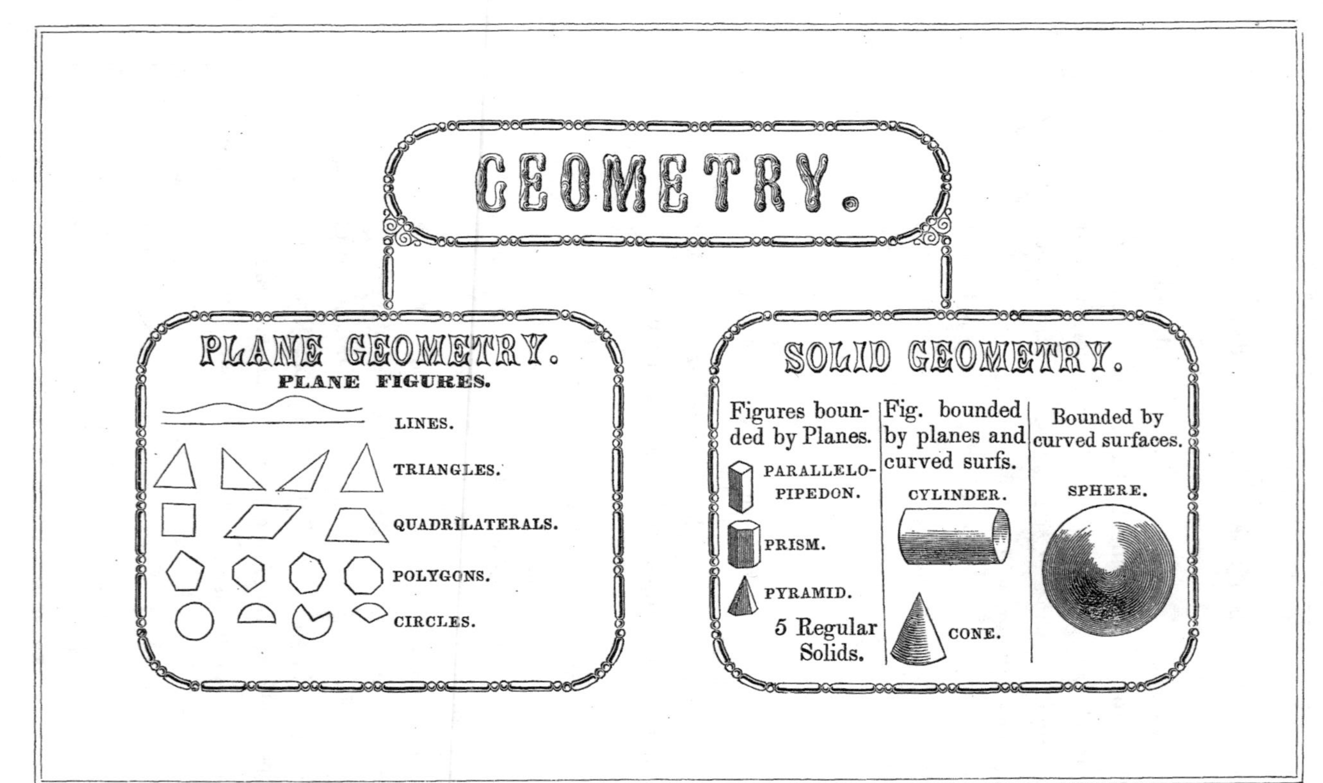
GEOMETRY.
PLANE GEOMETRY.
PLANE FIGURES.
LINES.
TRIANGLES.
QUADRILATERALS.
POLYGONS.
CIRCLES.
SOLID GEOMETRY.
Figures bounded by Planes.
PARALLELOPIPEDON.
PRISM.
PYRAMID.
5 Regular Solids.
Fig. bounded by planes and curved surfs.
CYLINDER.
CONE.
Bounded by curved surfaces.
SPHERE.

# CHAPTER III.

GEOMETRY DEFINED—THINGS OF WHICH IT TREATS—COMPARISON AND PROPERTIES OF FIGURES — DEMONSTRATION — PROPORTION — SUGGESTIONS FOR TEACHING.

## GEOMETRY.

§ 237. GEOMETRY treats of space, and compares portions of space with each other, for the purpose of pointing out their properties and mutual relations. The science consists in the development of all the laws relating to space, and is made up of the processes and rules, by means of which portions of space can be best compared with each other. The truths of Geometry are a series of dependent propositions, and may be divided into three classes: Geometry. Its science. Its truths. Of three kinds.

1st. Truths implied in the definitions, viz. that things do exist, or may exist, corresponding to the words defined. For example: when we say, "A quadrilateral is a rectilinear figure having four sides," we imply the existence of such a figure. 1st. Those implied in the definitions.

2d. Self-evident, or intuitive truths, embodied in the axioms; and, 2d. Axioms.

3d. Truths inferred from the definitions and 3d. Demonstrative truths. axioms, called Demonstrative Truths. We say that a truth or proposition is proved or demonstrated, when, by a course of reasoning, it is shown to be included under some other truth or proposition, previously known, and from which is said to *follow;* hence,

When demonstrated.

Demonstration. A DEMONSTRATION is a series of logical arguments, brought to a conclusion, in which the major premises are definitions, axioms, or propositions already established.

Subjects of Geometry. § 238. Before we can understand the proofs or demonstrations of Geometry, we must understand what that is to which demonstration is applicable: hence, the first thing necessary is to form a clear conception of space, the subject of all geometrical reasoning.*

Names of forms. The next step is to give names to particular forms or limited portions of space, and to define these names accurately. The definitions of these names are the definitions of Geometry, and the portions of space corresponding to them are called Figures, or Geometrical Magnitudes; of which there are three general classes:

Figures. Three kinds.

First. 1st. Lines;

Second. 2d. Surfaces;

Third. 3d. Solids.

* Sections 81 to 85.

§ 239. Lines embrace only one dimension of space, viz. length, without breadth or thickness. The extremities, or limits of a line, are called points.

Lines.

There are two general classes of lines—straight lines and curved lines. A straight line is one which lies in the same direction between any two of its points; and a curved line is one which constantly changes its direction at every point. There is but one kind of straight line, and that is fully characterized by the definition. From the definition we may infer the following axiom: "A straight line is the shortest distance between two points." There are many kinds of curves, of which the circumference of the circle is the simplest and the most easily described.

Two classes: Straight and Curved.

One kind of straight line.

many of curves.

§ 240. Surfaces embrace two dimensions of space, viz. length and breadth, but not thickness. Surfaces, like lines, are also divided into two general classes, viz. plane surfaces and curved surfaces.

Surfaces:

Plane and Curved.

A plane surface is that with which a straight line, any how placed, and having two points common with the surface, will coincide throughout its entire extent. Such a surface is perfectly even, and is commonly designated by the term "A plane." A large class of the figures

A plane surface;

Perfectly even.

Plane Figures. of Geometry are but portions of a plane, and all such are called plane figures.

A triangle, the most simple figure. § 241. A portion of a plane, bounded by three straight lines, is called a triangle, and this is the simplest of the plane figures. There are several kinds of triangles, differing from each other, however, only in the relative values of their sides and angles. For example: when the sides are all equal to each other, the triangle is called equilateral; when two of the sides are equal, it is called isosceles; and scalene, when the three sides are all unequal. If one of the angles is a right angle, the triangle is called a right-angled triangle.

Kinds of triangles.

Quadrilaterals. § 242. The next simplest class of plane figures comprises all those which are bounded by four straight lines, and are called quadrilaterals. There are several varieties of this class:

1st species. 1st. The mere quadrilateral, which has no mark, except that of having four sides ;

2d species. 2d. The trapezoid, which has two sides parallel and two not parallel;

3d species. 3d. The parallelogram, which has its opposite sides parallel and its angles oblique;

4th species. 4th. The rectangle, which has all its angles right angles and its opposite sides parallel; and,

5th. The square, which has its four sides equal to each other, each to each, and its four angles right angles. 5th species.

§ 243. Plane figures, bounded by straight lines, having a number of sides greater than four, take names corresponding to the number of sides, viz. Pentagons, Hexagons, Heptagons, &c. Other Polygons.

§ 244. A portion of a plane bounded by a curved line, all the points of which are equally distant from a certain point within called the centre, is called a circle, and the bounding line is called the circumference. This is the only curve usually treated of in Elementary Geometry. Circles: the circumference.

§ 245. A curved surface, like a plane, embraces the two dimensions of length and breadth. It is not even, like the plane, throughout its whole extent, and therefore a straight line may have two points in common, and yet not coincide with it. The surface of the cone, of the sphere, and cylinder, are the curved surfaces treated of in Elementary Geometry. Curved Surfaces: their properties.

§ 246. A solid is a portion of space, combining the three dimensions of length, breadth, and thickness. Solids are divided into three classes: Solids. Three classes.

1st class. 1st. Those bounded by planes;

2d class. 2d. Those bounded by plane and curved surfaces; and,

3d class. 3d. Those bounded only by curved surfaces.

What figures fall in each class. The first class embraces the pyramid and prism with their several varieties; the second class embraces the cylinder and cone; and the third class the sphere, together with others not generally treated of in Elementary Geometry.

Magnitudes named. § 247. We have now named all the geometrical magnitudes treated of in elementary Geometry. What they are. They are merely limited portions of space, and do not, *necessarily*, involve the idea of matter. A sphere. A sphere, for example, fulfils all the conditions imposed by its definitions, without any reference to what may be within the space enclosed by its surface. Need not be material. That space may be occupied by lead, iron, or air, or may be a vacuum, without at all changing the nature of the sphere, as a geometrical magnitude.

It should be observed that the boundary or limit of a geometrical magnitude, is another geometrical magnitude, having one dimension less. Boundaries of solids. For example: the boundary or limit of a solid, which has three dimensions, is always a surface which has but two: the limits or boundaries of Examples.

all surfaces are lines, straight or curved; and the extremities or limits of lines are points.

§ 248. We have now named and shown the nature of the things which are the subjects of Elementary Geometry. The science of Geometry is a collection of those connected processes by which we determine the measures, properties, and relations of these magnitudes.

Subjects named.

Science of Geometry.

COMPARISON OF FIGURES WITH UNITS OF MEASURE.

§ 249. We have seen that the term measure implies a comparison of the thing measured with some known thing of the same kind, regarded as a standard; and that such standard is called the unit of measure.* The unit of measure for lines must, therefore, be a line of a known length: a foot, a yard, a rod, a mile, or any other known unit. For surfaces, it is a square constructed on the linear unit as a side: that is, a square foot, a square yard, a square rod, a square mile; that is, *a square described on any known unit of length.*

Measure.

Unit of measure For Lines, A Line.

For Surfaces, A Square.

The unit of measure, for solidity, is a solid, and therefore has three dimensions. It is a cube

For Solids, A Cube.

* Section 94.

constructed on a linear unit as an edge, or on the superficial unit as a base. It is, therefore, a cubic foot, a cubic yard, a cubic rod, &c. Hence, there are three units of measure, each differing in kind from the other two, viz. a known length for the measurement of lines; a known square for the measurement of surfaces; and a known cube for the measurement of solids. The measure or contents of any magnitude, belonging to either class, is ascertained by finding how many times that magnitude contains its unit of measure.

Three units of measure: A Line, A Square, A Cube. Contents: how ascertained.

§ 250. There is yet another class of magnitudes with which Geometry is conversant, called Angles. They are not, however, elementary magnitudes, but arise from the relative positions of those already described. The unit of this class is the right angle; and with this as a standard, all other angles are compared

Angles: Their unit.

§ 251. We have dwelt with much detail on the unit of measure, because it furnishes the only basis of estimating quantity. The conception of number and space merely opens to the intellectual vision an unmeasured field of investigation and thought, as the ascent to the summit of a mountain presents to the eye a

Importance of the unit of measure:

wide and unsurveyed horizon. To ascertain the height of the point of view, the diameter of the surrounding circular area and the distance to any point which may be seen, some standard or unity must be known, and its value distinctly apprehended. So, also, number and space, which at first fill the mind with vague and indefinite conceptions, are to be finally measured by units of ascertained value.

Space indefinite without it: and always measured by it.

§ 252. It is found, on careful analysis, that every number may be referred to the unit one, as a standard, and when the signification of the term ONE is clearly apprehended, that any number, whether large or small, whether integer or fractional, may be deduced from the standard by an easy and known process.

Every number may be referred to the unit one.

In space, also, which is indefinite in extent, and exactly similar in all its parts, the faculties of the mind have established ideal boundaries. These boundaries give rise to the geometrical magnitudes, each of which has its own unit of measure; and by these simple contrivances, we measure space, even to the stars, as with a yard-stick.

Space: Its ideal boundaries.

§ 253. We have, thus far, not alluded to the difficulty of *determining* the exact length of that

Conception of the unit of measure: which we regard as a standard. We are presented with a given length, and told that it is called a foot or a yard, and this being usually done at a period of life when the mind is satisfied with mere facts, we adopt the conception of a distance corresponding to a name, and then by multiplying and dividing that distance we are enabled to apprehend other distances. But this by no means answers the inquiry, What is the standard for measurement?

At first, a mere impression.

How determined.

Under the supposition that the laws of physical nature operate uniformly, the unit of measure in England and the United States has been fixed by ascertaining the length of a pendulum which will vibrate seconds, and to this length the Imperial yard, which we have also adopted as a standard, is referred. Hence, the unit of measure is referred to a natural standard, viz. to the distance between the axis of suspension and the centre of oscillation of a pendulum which shall vibrate seconds in vacuo, in London, at the level of the sea. This distance is declared to be **39.1393** *imperial inches;* that is, **3** imperial feet and **3.1393** inches. Thus, the determination of the unit of length demands the application of the most abstruse science, combined with accurate observation and delicate experiment.

What it is.

Its length.

Difficulties of determining it.

Could this distance, or unit, have been exactly

ascertained before the measures of the world were fixed, and in general use, it would have afforded a standard at once certain and convenient, and all distances would then have been expressed in numbers arising from its multiplication or exact division. But as the measures of the world (and consequently their units) were fixed antecedently to the determination of this distance, it was expressed in measures already known; and hence, instead of being represented by 1, which had already been appropriated to the foot, it was expressed in terms of the foot, viz. 39.1393 inches, and this is now the standard to which all units of measure are referred.

What should have been called one.

Other numbers derived from it.

Why it is not so.

What now represents it.

§ 254. The unit of measure is not only important as affording a basis for all measurement, but is also the element from which we deduce the unit of weight. The weight of 27.7015 cubic inches of distilled water is taken as the standard, weighing exactly one pound avoirdupois, and this quantity of water is determined from the unit of length; that is, the determination of it reaches back to the length of a pendulum which will vibrate seconds in the latitude of London.

Unit of measure the basis of the unit of weight.

What it is.

§ 255. Two geometrical figures are said to be equivalent, when they contain the same unit of

Equivalent figures.

measure an equal number of times. Two figures are said to be equal when they can be so applied to each other as to coincide throughout their whole extent. Hence, equivalency refers to measure, and equality to coincidence. Indeed, coincidence is the only test of geometrical equality. All equal figures are of course equivalent, though equivalent figures are by no means equal. Equality is equivalency, with the further mark of coincidence.

Equal figures.

Equivalency: Equality.

Their difference.

## PROPERTIES OF FIGURES.

Property of figures.

§ 256. A property of a figure is a mark common to all figures of the same class. For example: if the class be "Quadrilateral," there are two very obvious properties, common to all quadrilaterals, besides the one which characterizes the figure, and by which its name is defined, viz. that it has four angles, and that it may be divided into two triangles. If the class be "Parallelogram," there are several properties common to all parallelograms, and which are subjects of proof; such as, that the opposite sides and angles are equal; the diagonals divide each other into equal parts, &c. If the class be "Triangle," there are many properties common to all triangles, besides the characteristic that

Quadrilaterals.

Parallelogram.

Triangle:

they have three sides. If the class be a particular kind of triangle, such as the equilateral, isosceles, or right-angled triangle, then each class has particular properties, common to every individual of the class, but not common to the other classes. It is important, however, to remark, that every property which belongs to "triangle," regarded as a genus, will appertain to every species or class of triangle; and universally, every property which belongs to a genus will belong to every species under it; and every property which belongs to a species will belong to every class or subspecies under it; and every property which belongs to one of a subspecies or class will be common to every individual of the class. For example: "the square on the hypothenuse of a right-angled triangle is equivalent to the sum of the squares described on the other two sides," is a proposition equally true of every right-angled triangle: and "every straight line perpendicular to a chord, at the middle point, will pass through the centre," is equally true of all circles.

Equilateral, Isosceles. Right-angled.

Every property which belongs to a genus will be common to every species:

also to every subspecies, and to every individual.

Examples.

Circle.

MARKS OF WHAT MAY BE PROVED.

§ 257. The characteristic properties of every geometrical figure (that is, those properties with-

Characteristic properties.

out which the figures could not exist), are given in the definitions. How are we to arrive at all the other properties of these figures? The propositions implied in the definitions, viz. that things corresponding to the words defined do or may exist with the properties named; and the self-evident propositions or axioms, contain the only marks of what can be proved; and by a skilful combination of these marks we are able to discover and prove all that is discovered and proved in Geometry.

Marks: Of what may be proved. How extended.

Definitions and axioms, and propositions deduced from them, are major premises in each new demonstration; and the science is made up of the processes employed for bringing unforeseen cases under these known truths; or, in syllogistic language, for proving the minors necessary to complete the syllogisms. The marks being so few, and the inductions which furnish them so obvious and familiar, there would seem to be very little difficulty in the deductive processes which follow. The connecting together of several of these marks constitutes Deductions, or Trains of Reasoning; and hence, Geometry is a Deductive Science.

Major Premiss. The science: in what it consists.

Geometry, a Deductive Science.

DEMONSTRATION.

§ 258. As a first example, let us take the first proposition in Legendre's Geometry:

Proposition to be proved.

"*If a straight line meet another straight line, the sum of the two adjacent angles will be equal to two right angles.*"

Enunciation.

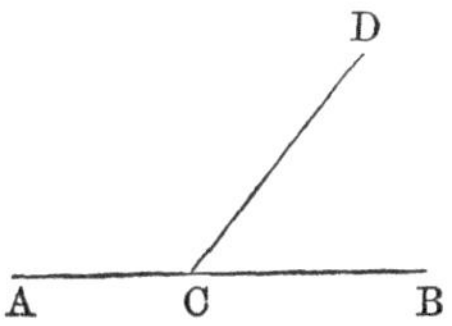

Let the straight line DC meet the straight line AB at the point C, then will the angle ACD plus the angle DCB be equal to two right angles.

Things necessary to prove it.

To prove this proposition, we need the definition of a right angle, viz.:

Definitions.

*When a straight line AB meets another straight line CD, so as to make the adjacent angles BAC and BAD equal to each other, each of those angles is called a* RIGHT ANGLE, *and the line AB is said to be* PERPENDICULAR *to CD.*

Axioms.

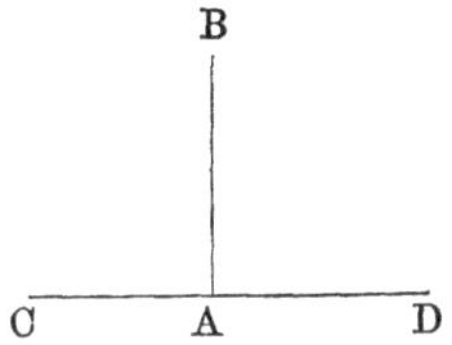

We shall also need the 2d, 3d, and 4th axioms, for inferring equality,* viz.:

Second.

2. Things which are equal to the same thing are equal to each other.

† Section 102.

Third. 3. A whole is equal to the sum of all its parts.

Fourth. 4. If equals be added to equals, the sums will be equal.

Now before these formulas or tests can be applied, it is necessary to suppose a straight line CE to be drawn perpendicular to AB at the point C: then by the definition of a right angle, the angle ACE will be equal to the angle ECB.

Line to be drawn.

Proof:

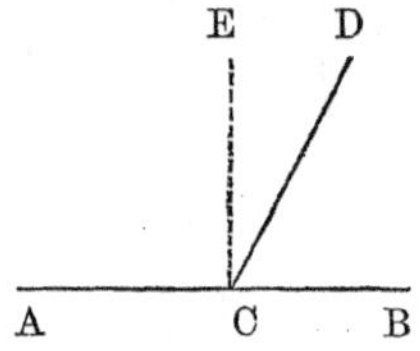

E D

A C B

By axiom 3rd, we have,

Continued: ACD equal to ACE plus ECD: to each of these equals add DCB; and by the 4th axiom we shall have,

ACD plus DCB equal to ACE plus ECD plus DCB; but by axiom 3rd,

ECD plus DCB equals ECB: therefore by axiom 2d,

ACD plus DCB equals ACE plus ECB.

But by the definition of a right angle,

Conclusion. ACE plus ECB equals two right angles: therefore, by the 2d axiom,

ACD plus DCB equals two right angles.

Its bases. It will be seen that the conclusiveness of the proof results,

First. 1st. From the definition, that ACE and ECB are equal to each other, and each is called a right-angle: consequently, their sum is equal to two right angles; and,

Second.

2dly. In showing, by means of the axioms, that ACD plus DCB equals ACE plus ECB; and then *inferring* from axiom 2d, that, ACD plus DCB equals two right angles.

Difficulties in the demonstrations.

§ 259. The difficulty in the geometrical reasoning consists mainly in showing that the proposition to be proved contains the marks which prove it. To accomplish this, it is frequently necessary to draw many auxiliary lines, forming new figures and angles, which can be shown to possess marks of these marks, and which thus become connecting links between the known and the unknown truths. Indeed, most of the skill and ingenuity exhibited in the geometrical processes are employed in the use of these auxiliary means. The example above affords an illustration. We were unable to show that the sum of the two angles possessed the mark of being equal to two right angles, until we had drawn a perpendicular, or supposed one drawn, at the point where the given lines intersect. That being done, the two right angles ACE and ECB were formed, which enabled us to *compare* the sum of the angle ACD and DCB with two right angles, and thus we *proved* the proposition.

Auxiliaries necessary.

Connecting Links.

How used.

Conclusion.

Proposition. § 260. As a second example, let us take the following proposition:

Enunciation. *If two straight lines meet each other, the opposite or vertical angles will be equal.*

Diagram. Let the straight line AB meet the straight line ED at the point C: then will the angle ACD be equal to the opposite angle ECB; and the angle ACE equal to the angle DCB.

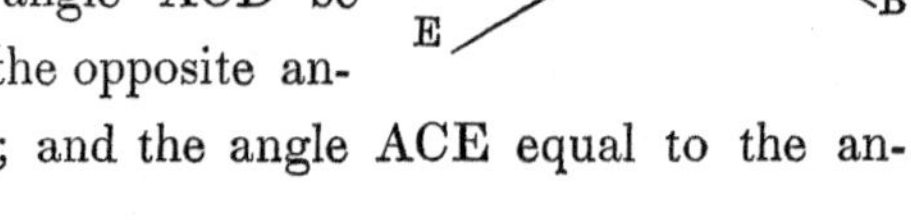


Principles necessary. To prove this proposition, we need the last proposition, and also the 2d and 5th axioms, viz.:

"If a straight line meet another straight line, the sum of the two adjacent angles will be equal to two right angles."

Axioms. "Things which are equal to the same thing are equal to each other."

"If equals be taken from equals, the remainders will be equal."

Now, since the straight line AC meets the straight line ED at the point C, we have,

Proof. ACD plus ACE equal to two right angles.

And since the straight line DC meets the straight line AB, we have,

ACD plus DCB equal to two right angles: hence, by the second axiom,

ACD plus ACE equals ACD plus DCB: taking from each the common angle ACD, we know from the fifth axiom that the remainders will be equal; that is, the angle ACE equal to the opposite or vertical angle DCB. Conclusion.

§ 261. The two demonstrations given above combine all the processes of proof employed in every demonstration of the same class. When any new truth is to be proved, the known tests of truth are gradually extended to auxiliary quantities having a more intimate connection with such new truth than existed between it and the known tests, until finally, the known tests, through a series of links, become applicable to the final truth to be established: the intermediate processes, as it were, bridging over the space between the known tests and the final truth to be proved. Demonstrations general. Use of auxiliary quantities.

§ 262. There are two classes of demonstrations, quite different from each other, in some respects, although the same processes of argumentation are employed in both, and although the conclusions in both are subjected to the same logical tests. They are called Direct, or Positive Demonstration, and Negative Demonstration, or the Reductio ad Absurdum. Direct demonstration. Negative, or Reductio ad Absurdum.

Difference. Direct Demonstration. Negative Demonstration. Conclusion: With what compared. Determines whether the hypothesis is true or false.

§ 263. The main differences in the two methods are these: The method of direct demonstration rests its arguments on known and admitted truths, and shows by logical processes that the proposition can be brought under some previous definition, axiom, or proposition: while the negative demonstration rests its arguments on an hypothesis, combines this with known propositions, and deduces a conclusion by processes strictly logical. Now if the conclusion so deduced agrees with any known truth, we infer that the hypothesis, (which was the only link in the chain not previously known), was true; but if the conclusion be excluded from the truths previously established; that is, if it be opposed to any one of them, then it follows that the hypothesis, being contradictory to such truth, must be false. In the negative demonstration, therefore, the *conclusion* is compared with the truths known antecedently to the proposition in question: if it agrees with any one of them, the hypothesis is correct; if it disagrees with any one of them, the hypothesis is false.

Proof by Negative Demonstration.

§ 264. We will give for an illustration of this method, Proposition XVII. of the First Book of Legendre: "When two right-angled triangles have the hypothenuse and a side of the one equal to the hypothenuse and a side of the other, each to each, the remaining parts will be equal, each to each, and the triangles themselves will be equal." Enunciation.

In the two right-angled triangles BAC and EDF (see next figure), let the hypothenuse AC be equal to DF, the side BA to the side ED: then will the side BC be equal to EF, the angle A to the angle D, and the angle C to the angle F. To prove this proposition, we need the following, which have been before proved; viz.: Enunciation by the figure.

Prop. X. (of Legendre). "When two triangles have the three sides of the one equal to the three sides of the other, each to each, the three angles will also be equal, each to each, and the triangles themselves will be equal." Previous truths necessary.

Prop. V. "When two triangles have two sides and the included angle of the one, equal to two sides and the included angle of the other, each to each, the two triangles will be equal." Proposition

Axiom I. "Things which are equal to the same thing, are equal to each other." Axioms.

Axiom X. (of Legendre). "All right angles are equal to each other."

Prop. XV. "If from a point without a straight line, a perpendicular be let fall on the line, and oblique lines be drawn to different points, Proposition.

1st. "The perpendicular will be shorter than any oblique line;

2d. "Of two oblique lines, drawn at pleasure, that which is farther from the perpendicular will be the longer."

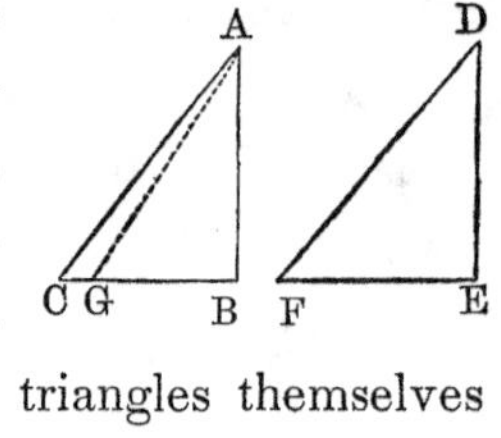


Beginning of the demonstration. Now the two sides BC and EF are either equal or unequal. If they are equal, then by Prop. X. the remaining parts of the two triangles are also equal, and the triangles themselves are equal. If the two sides are unequal, one of them must be greater than the other: suppose BC to be the greater.

Construction of the figure. On the greater side BC take a part BG, equal to EF, and draw AG. Then, in the two triangles BAG and DEF the angle B is equal to the angle E, by axiom X (Legendre), both being right angles. The side AB is equal to the side DE, and by hypothesis the side BG is equal to the side EF: then it follows from Prop. V. that the side AG is equal to the side DF. But the side

Demonstration. DF is equal to the side AC: hence, by axiom I, the side AG is equal to AC. But the line AG cannot be equal to the line AC, having been shown to be less than it by Prop. XV.: hence,

Conclusion. the conclusion contradicts a known truth, and is therefore false; consequently, the supposition (on which the conclusion rests), that BC and EF are unequal, is also false; therefore, they are equal.

§ 265. It is often supposed, though erroneously, that the Negative Demonstration, or the demonstration involving the "reductio ad absurdum," is less conclusive and satisfactory than direct or positive demonstration. This impression is simply the result of a want of proper analysis. For example: in the demonstration just given, it was proved that the two sides BC and EF cannot be *unequal;* for, such a supposition, in a logical argumentation, resulted in a conclusion directly opposed to a known truth; and as equality and inequality are the only general conditions of relation between two quantities, it follows that if they do not fulfil the one, they *must* the other. In both kinds of demonstration, the premises and conclusion agree; that is, they are both true, or both false; and the reasoning or argument in both is supposed to be strictly logical.

Negative Demonstration: Conclusive. Reasons. Conclusion corresponds to, or is opposed to known truth. Agreement.

In the direct demonstration, the premises are known, being antecedent truths; and hence, the conclusion is true. In the negative demonstration, *one element* is assumed, and the conclusion is then compared with truths previously established. If the conclusion is found to agree with any one of these, we infer that the hypothesis or assumed element is true; if it contradicts any one of these truths, we infer that

Differences in the two kinds. When the hypothesis is true.

When false. the assumed element is false, and hence that its opposite is true.

Measured: its signification. § 266. Having explained the meaning of the term measured, as applied to a geometrical magnitude, viz. that it implies the comparison of a magnitude with its unit of measure; and having also explained the signification of the word Property, and the processes of reasoning by which, in all figures, properties not before noticed are inferred from those that are known; we shall now add a few remarks on the relations of the geometrical figures, and the methods of comparing them with each other.

General Remarks.

## PROPORTION OF FIGURES.

Proportion: § 267. Proportion is the relation which one geometrical magnitude bears to another of the same kind, with respect to its being greater or less. The two magnitudes so compared are called terms, and the measure of the proportion is the quotient which arises from dividing the second term by the first, and is called their Ratio. Only quantities of the same kind can be compared together, and it follows from the nature of the relation that the quotient or ratio of any two terms will be an abstract number, whether the terms themselves be abstract or concrete.

Its measure.

Ratio.

Quantities of the same kind compared.

§ 268. The term Proportion is defined by most authors, "An equality of ratios between four numbers or quantities, compared together two and two." A proportion certainly arises from such a comparison: thus, if

Proportion: how defined.

$$\frac{B}{A} = \frac{D}{C}; \text{ then,}$$

Example.

$$A : B : : C : D$$

is a proportion.

But if we have two quantities A and B, which may change their values, and are, at the same time, so connected together that one of them shall increase or decrease just as many times as the other, their ratio will not be altered by such changes; and the two quantities are then said to be in proportion, or proportional.

True definition.

Two proportional quantities.

Thus, if A be increased three times and B three times, then,

$$\frac{3\,B}{3\,A} = \frac{A}{B};$$

that is, 3 A and 3 B bear to each other the same proportion as A and B. Science needed a general term to express this relation between *two* quantities which change their values, without altering their quotient, and the term "proportional," or "in proportion," is employed for that purpose.

Term needed.

How used.

Reasons for modification.

As some apology for the modification of the definition of proportion, which has been so long accepted, it may be proper to state that the term has been used by the best authors in the exact sense here attributed to it. In the definition of the second law of motion, we have, "Motion, or change of motion, is proportional to the force impressed;"* and again, "The inertia of a body is proportioned to its weight."† Similar examples may be multiplied to any extent. Indeed, there is a symbol or character to express the relation between two quantities, when they undergo changes of value, without altering their ratio. That character is $\propto$, and is read "proportional to." Thus, if we have two quantities denoted by A and B, written,

Use of the term.

Symbol used to represent proportion.

Example.

$$A \propto B,$$

the expression is read, "A proportional to B."

Another kind of proportion.

§ 269. There is yet another kind of relation which may exist between two quantities A and B, which it is very important to consider and understand. Suppose the quantities to be so connected with each other, that when the first is increased according to any law of change, the second shall decrease according to the same law; and the reverse.

* Olmsted's Mechanics, p. 28. † Ibid. p. 23.

First Example.

For example: the area of a rectangle is equal to the product of its base and altitude. Then, in the rectangle ABCD, we have

$$\text{Area} = AB \times BC.$$

Second Example.

Take a second rectangle EFGH, having a longer base EF, and a less altitude FG, but such that it shall have an equal area with the first: then we shall have

$$\text{Area} = EF \times FG.$$

Now since the areas are equal, we shall have

Equation.

$$AB \times BC = EF \times FG;$$

and by resolving the terms of this equation into a proportion, we shall have

Proportion.

$$AB : EF :: FG : BC.$$

Relations of the quantities:

It is plain that the sides of the rectangle ABCD may be so changed in value as to become the sides of the rectangle EFGH, and that while they are undergoing this change, AB will increase and BC diminish. The change in the values of these quantities will therefore take place according to a fixed law: that is, one will be diminished as many times as the other is increased, since their product is constantly equal to the area of the rectangle EFGH.

Expressed by letters.

Denote the side AB by $x$ and BC by $y$, and the area of the rectangle EFGH, which is known, by $a$; then

$$xy = a;$$

and when the product of two varying quantities is constantly equal to a known quantity, the two quantities are said to be *Reciprocally* or *Inversely* proportional; thus $x$ and $y$ are said to be inversely proportional to each other. If we divide 1 by each member of the above equation, we shall have

Reciprocal or Inverse Proportion.

$$\frac{1}{xy} = \frac{1}{a};$$

Reductions of the Equations.

and by multiplying both members by $x$, we shall have

$$\frac{1}{y} = \frac{x}{a};$$

and then by dividing both numbers by $x$, we have

Final form.

$$\frac{\frac{1}{y}}{x} = \frac{1}{a};$$

that is, the ratio of $x$ to $\frac{1}{y}$ is constantly equal to $\frac{1}{a}$; that is, equal to the same quantity, however $x$ or

$y$ may vary; for, $a$ and consequently $\frac{1}{a}$ does not change. Hence,

*Two quantities, which may change their values, are reciprocally or inversely proportional, when one is proportional to unity divided by the other, and then their product remains constant.*

Inverse Proportion defined.

We express this reciprocal or inverse relation thus:

$$A \propto \frac{1}{B}.$$

A is said to be inversely proportional to B: the symbols also express that A is directly proportional to $\frac{1}{B}$. If we have

How expressed.

$$A \propto \frac{B}{C},$$

we say, that A is directly proportional to B, and inversely proportional to C.

Generally, how read.

The terms Direct, Inverse or Reciprocal, apply to the *nature* of the *proportion*, and not to the Ratio, which is always a mere quotient and the measure of proportion. The term Direct applies to all proportions in which the terms increase or decrease together; and the term Inverse or Reciprocal to those in which one term increases as the other decreases. They cannot, therefore, properly be applied to ratio without changing entirely its signification and definition.

Direct and Inverse, terms not applicable to Ratio.

## COMPARISON OF FIGURES.

Geometrical magnitudes compared.

§ 270. In comparing geometrical magnitudes, by means of their quotient, it is not the quotient alone which we consider. The comparison implies a general relation of the magnitudes, which is measured by the Ratio. For example: we say that "Similar triangles are to each other as the squares of their homologous sides." What do we mean by that? Just this:

Example.

Formula of Comparison.

That the area of a triangle
Is to the area of a similar triangle
As the area of a square described on a side of the first,
To the area of a square described on an homologous side of the second.

Changes of value: how affected

Thus, we see that every term of such a proportion is in fact a surface, and that the area of a triangle increases or decreases much faster than its sides; that is, if we double each side of a triangle, the area will be four times as great: if we multiply each side by three, the area will be nine times as great; or if we divide each side by two, we diminish the area four times, and so on. Again,

Results.

Circles compared.

The area of one circle
Is to the area of another circle,
As a square described on the diameter of the first

To a square described on the diameter of the second.

Hence, if we double the diameter of a circle, the area of the circle whose diameter is so increased will be four times as great: if we multiply the diameter by three, the area will be nine times as great; and similarly if we divide the diameter. How their areas change. Principle general.

§ 271. In comparing solids together, the same general principles obtain. Similar solids are to each other as the cubes described on their homologous or corresponding sides. That is, Comparison of solids.

A prism
Is to a similar prism,
As a cube described on a side of the first,
Is to a cube described on an homologous side of the second. Formula.

Hence, if the sides of a prism be doubled, the solid contents will be increased eight-fold. Again, How the solidities change.

A sphere
Is to a sphere,
As a cube described on the diameter of the first,
Is to a cube described on a diameter of the second. Sphere:

Hence, if the diameter of a sphere be doubled, its solid contents will be increased eight-fold; if the diameter be multiplied by three, the solid How its solidity changes.

contents will be increased twenty-seven fold: if the diameter be multiplied by four, the solid contents will be increased sixty-four fold; the solid contents increasing as the cubes of the numbers 1, 2, 3, 4, &c.

Ratio: an abstract number. When having a fixed value. How varying solids are compared.

§ 272. The relation or ratio of two magnitudes to each other, may be, and indeed is, expressed by an abstract number. This number has a fixed value so long as we do not introduce a change in the volumes of the solids; but if we wish to express their ratio under the supposition that their volumes may change according to fixed laws (that is, so that the solids shall continue similar), we then compare them with similar figures described on their homologous or corresponding sides; or, what is the same thing, take into account the corresponding changes which take place in the abstract numbers that express their volumes.

## RECAPITULATION.

General outline.

§ 273. We have now completed a general outline of the science of Geometry, and what has been said may be recapitulated under the following heads. It has been shown,

Geometry;

1st. That Geometry is conversant about space,

or those limited portions of space which are called Geometrical Magnitudes. to what it relates.

2d. That the geometrical magnitudes embrace three species of figures:

1st. Lines—straight and curved; Lines.

2d. Surfaces—plane and curved; Surfaces.

3d. Solids—bounded either by plane surfaces or curved, or both; and, Solids.

4th. Angles, arising from the positions of lines and planes, and by which they are bounded. Angles.

3d. That the science of Geometry is made up of those processes by means of which all the properties of these magnitudes are examined and developed, and that the results arrived at constitute the truths of Geometry. Science: how made up.

4th. That the truths of Geometry may be divided into three classes: Truths: three classes.

1st. Those implied in the definitions, viz. that things exist corresponding to certain words defined; First class.

2d. Intuitive or self-evident truths embodied in the axioms; Second.

3d. Truths deduced (that is, inferred) from the definitions and axioms, called Demonstrative Truths. Third.

5th. That the examination of the properties of the geometrical magnitudes has reference, Geometrical magnitudes.

Comparison. 1st. To their comparison with a standard or unit of measure;

Properties. 2d. To the discovery of properties belonging to an individual figure, and yet common to the entire class to which such figure belongs;

Proportion. 3d. To the comparison, with respect to magnitude, of figures of the same species with each other; viz. lines with lines, surfaces with surfaces, and solids with solids.

### SUGGESTIONS FOR THOSE WHO TEACH GEOMETRY.

Suggestions. First. 1. Be sure that your pupils have a clear apprehension of space, and of the notion that Geometry is conversant about space only.

Second. 2. Be sure that they understand the signification of the terms, lines, surfaces, and solids, and that these names indicate certain portions of space corresponding to them.

Third. 3. See that they understand the distinction between a straight line and a curve; between a plane surface and a curved surface; between a solid bounded by planes and a solid bounded by curved surfaces.

Fourth. 4. Be careful to have them note the characteristics of the different species of plane figures, such as triangles, quadrilaterals, pentagons, hexagons, &c.; and then the characteristic of each class or subspecies, so that the name shall recall, at once, the characteristic properties of each figure.

5. Be careful, also, to have them note the characteristic differences of the solids. Let them often name and distinguish those which are bounded by planes, those bounded by plane and curved surfaces, and those bounded by curved surfaces only. Regarding Solids as a genus, let them give the species and subspecies into which the solid bodies may be divided. Fifth.

6. Having thus made them familiar with the things which are the subjects of the reasoning, explain carefully the nature of the definitions; then of the axioms, the grounds of our belief in them, and the information from which those self-evident truths are inferred. Sixth.

7. Then explain to them, that the definitions and axioms are the basis of all geometrical reasoning: that every proposition must be deduced from them, and that they afford the tests of all the truths which the reasonings establish. Seventh.

8. Let every figure, used in a demonstration, be accurately drawn, by the pupil himself, on a blackboard. This will establish a connection between the eye and the hand, and give, at the same time, a clear perception of the figure and a distinct apprehension of the relations of its parts. Eighth.

Ninth. 9. Let the pupil, in every demonstration, first enunciate, in general terms, that is, without the aid of a diagram, or any reference to one, the proposition to be proved; and then state the principles previously established, which are to be employed in making out the proof.

Tenth. 10. When in the course of a demonstration, any truth is inferred from its connection with one before known, let the truth so referred to be fully and accurately stated, even though the number of the proposition in which it is proved, be also required. This is deemed important.

Eleventh. 11. Let the pupil be made to understand that a demonstration is but a series of logical arguments arising from comparison, and that the result of every comparison, in respect to quantity, contains the mark either of equality or inequality.

Twelfth. 12. Let the distinction between a positive and negative demonstration be fully explained and clearly apprehended.

Thirteenth. 13. In the comparison of quantities with each other, great care should be taken to impress the fact that proportion exists only between quantities of the same kind, and that ratio is the measure of proportion.

Fourteenth. 14. Do not fail to give much importance to the *kind* of quantity under consideration. Let the question be often put, What kind of quantity are you considering? Is it a line, a surface, or a solid? And what kind of a line, surface, or solid? Fourteenth.

15. In all cases of measurement, the unit of measure should receive special attention. If lines are measured, or compared by means of a common unit, see that the pupil perceives that unit clearly, and apprehends distinctly its relations to the lines which it measures. In surfaces, take much pains to mark out on the blackboard the particular square which forms the unit of measure, and write unit, or unit of measure, over it. So in the measurement of solidity, let the unit or measuring cube be exhibited, and the conception of its size clearly formed in the mind; and then impress the important fact, that, *all measurement consists in merely comparing a unit of measure with the quantity measured; and that the number which expresses the ratio is the numerical expression for that measure.* Fifteenth.

16. Be careful to explain the difference of the terms *Equal* and *Equivalent,* and never permit the pupil to use them as synonymous. An accurate use of words leads to nice discriminations of thought. Sixteenth.

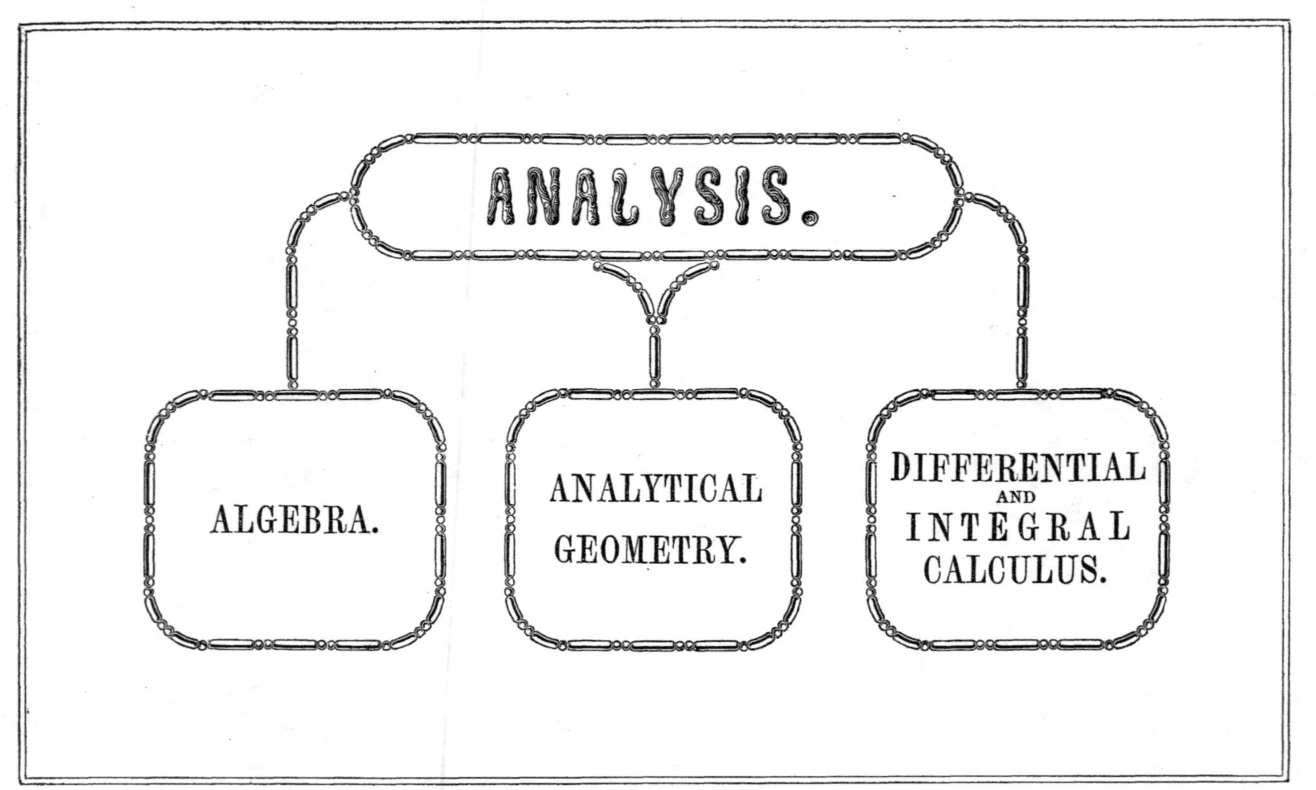
ANALYSIS.
ALGEBRA.
ANALYTICAL GEOMETRY.
DIFFERENTIAL AND INTEGRAL CALCULUS.

# CHAPTER IV.

ANALYSIS—ALGEBRA—ANALYTICAL GEOMETRY—DIFFERENTIAL AND INTEGRAL CALCULUS.

## ANALYSIS.

§ 274. ANALYSIS is a general term, embracing that entire portion of mathematical science in which the quantities considered are represented by letters of the alphabet, and the operations to be performed on them are indicated by signs.

Analysis defined.

§ 275. We have seen that all numbers must be numbers of something;* for, there is no such thing as a number without a basis: that is, every number must be based on the abstract unit one, or on some unit denominated. But although numbers must be numbers of *something*, yet they may be numbers of *any thing*, for the unit may be whatever we choose to make it.

Numbers must be of things; but may be of many kind of things.

* Section 112.

All quantity consists of parts.

§ 276. All quantity consists of parts, which can be numbered exactly or approximatively, and, in this respect, possesses all the properties of number. Propositions, therefore, concerning numbers, have the remarkable peculiarity, that they are propositions concerning all quantities whatever. That half of six is three, is equally true, whatever the word six may represent, whether six abstract units, six men, or six triangles. Analysis extends the generalization still further. A number represents, or stands for, that particular number of things of the same kind, without reference to the *nature* of the thing; but an analytical symbol does more, for it may stand for *all numbers,* or for all quantities which numbers represent, or even for quantities which cannot be exactly expressed numerically.

Propositions in regard to number apply also to quantity.

Algebraic symbols more general.

Any thing conceived may be divided.

As soon as we conceive of a thing we may conceive it divided into equal parts, and may represent either or all of those parts by $a$ or $x$, or may, if we please, denote the thing itself by $a$ or $x$, without any reference to its being divided into parts.

Each figure stands for a class.

§ 277. In Geometry, each geometrical figure stands for a class; and when we have demonstrated a property of a figure, that property is considered as proved for every figure of the class.

For example: when we prove that the square described on the hypothenuse of a right-angled triangle is equivalent to the sum of the squares described on the other two sides, we demonstrate the fact for all right-angled triangles. But in analysis, all numbers, all lines, all surfaces, all solids, may be denoted by a single symbol, $a$ or $x$. Hence, all truths inferred by means of these symbols are true of all things whatever, and not like those of number and geometry, true only of particular classes of things. It is, therefore, not surprising, that the symbols of analysis do not excite in our minds the ideas of particular things. The mere written characters, $a$, $b$, $c$, $d$, $x$, $y$, $z$, serve as the representatives of things in general, whether abstract or concrete, whether known or unknown, whether finite or infinite.

Example.

In analysis the symbols stand for things of all classes.

Hence, the truths inferred are general.

§ 278. In the uses which we make of these symbols, and the processes of reasoning carried on by means of them, the mind insensibly comes to regard them as *things*, and not as mere signs; and we constantly predicate of them the properties of things in general, without pausing to inquire what kind of thing is implied. Thus, we define an equation to be a proposition in which equality is predicated of one thing as compared with another. For example:

Symbols come to be regarded as things.

Example.

The equation.

$$a + c = x,$$

What axioms necessary to its solution.

is an equation, because $x$ is declared to be equal to the sum of $a$ and $c$. In the solution of equations, we employ the axioms, "If equals be added to equals, the sums will be equal;" and, "If equals be taken from equals, the remainders will be equal." Now, these axioms do not express qualities of language, but properties of quantity. Hence, all inferences in mathematical science, deduced through the instrumentality of symbols, whether Arithmetical, Geometrical, or Analytical, must be regarded as concerning quantity, and not symbols.

They express qualities of things.

Hence, inferences relate to things.

Quantity need not always be present to the mind.

As analytical symbols are the representatives of quantity in general, there is no necessity of keeping the idea of *quantity* continually alive in the mind; and the processes of thought may, without danger, be allowed to rest on the symbols themselves, and therefore, become to that extent, merely mechanical. But, when we look back and see on what the reasoning is based, and how the processes have been conducted, we shall find that every step was taken on the supposition that we were actually dealing with things, and not symbols; and that, without this understanding of the language, the whole system is without signification, and fails.

The reasoning is all based on the supposition of quantity.

§ 279. There are three principal branches of Analysis : Three branches :

1st. Algebra. Algebra,

2d. Analytical Geometry. Analytical Geometry,

3d. Differential and Integral Calculus. Calculus.

## ALGEBRA.

§ 280. Algebra is, in fact, a species of universal Arithmetic, in which letters and signs are employed to abridge and generalize all processes involving numbers. It is divided into two parts, corresponding to the science and art of Arithmetic : Algebra: Universal Arithmetic, Two parts:

1st. That which has for its object the investigation of the properties of numbers, embracing all the processes of reasoning by which new properties are inferred from known ones ; and, First part.

2d. The solution of all problems or questions involving the determination of certain numbers which are unknown, from their connection with certain others which are known or given. Second part.

## ANALYTICAL GEOMETRY.

§ 281. Analytical Geometry examines the properties, measures, and relations of the geometrical magnitudes by means of the analytical Analytical Geometry. Its nature.

symbols. This branch of mathematical science originated with the illustrious Descartes, a celebrated French mathematician of the 17th century. He observed that the positions of points, the direction of lines, and the forms of surfaces, could be expressed by means of the algebraic symbols; and consequently, that every change, either in the position or extent of a geometrical magnitude, produced a corresponding change in certain symbols, by which such magnitude could be represented. As soon as it was found that, to every variety of position in points, direction in lines, or form of curves or surfaces, there corresponded certain analytical expressions (called their Equations), it followed, that if the processes were known by which these equations could be examined, the relation of their parts determined, and the laws according to which those parts vary, relative to one another, ascertained, then the corresponding changes in the geometrical magnitudes, thus represented, could be immediately inferred.

Descartes, the original founder of this science.

What he observed.

All position expressed by symbols.

The equation develops the properties of the magnitude.

Hence, it follows that every geometrical question can be solved, if we can resolve the corresponding algebraic equation; and the power over the geometrical magnitudes was extended just in proportion as the properties of quantity were brought to light by means of the Calculus. The

Power over the magnitude extended by the equation.

applications of this Calculus were soon made to the subjects of mechanics, astronomy, and indeed, in a greater or less degree, to all branches of natural philosophy; so that, at the present time, all the varieties of physical phenomena, with which the higher branches of the science are conversant, are found to answer to varieties determinable by the algebraic analysis.

To what subject applied.

Its present uses.

§ 282. Two classes of quantities, and consequently two sets of symbols, quite distinct from each other, enter into this Calculus; the one called *Constants,* which preserve a fixed or given value throughout the same discussion or investigation; and the other called Variables, which undergo certain changes of value, the laws of which are indicated by the algebraic expressions or equations into which they enter. Hence,

Quantities which enter into the Calculus.

Constants.

Variables.

Analytical Geometry may be defined as that branch of mathematical science, which examines, discusses, and develops the properties of geometrical magnitudes by noting the changes that take place in the algebraic symbols which represent them, the laws of change being determined by an algebraic equation or formula.

Analytical Geometry defined.

## DIFFERENTIAL AND INTEGRAL CALCULUS.

Quantities considered. § 283. In this branch of mathematical science, as in Analytical Geometry, two kinds of quantity are considered, viz. Variables and Constants; (Variables, Constants.) and consequently, two distinct sets of symbols are employed. (The Science.) The science consists of a series of processes which note the changes that take place in the value of the Variables. Those changes of value take place according to fixed laws established by algebraic formulas, and are indicated by certain *marks* drawn from the variable symbols, (Marks. Differential Coefficients.) called *Differential Coefficients.* By these marks we are enabled to trace out with the accuracy of exact science the most hidden properties of quantity, as well as the most general and minute laws which regulate its changes of value.

Analytical Geometry, and Calculus: § 284. It will be observed, that Analytical Geometry and the Differential and Integral Calculus treat of quantity regarded under the same general aspect, viz. as subject to changes or variations in magnitude according to laws indicated by algebraical formulas; (How they regard quantity:) and the quantities, whether variable or constant, are, in both cases, represented by the same algebraic symbols, viz. (by what represented.) the constants by the first, and the variables by the final letters of the alphabet. There is, however, this important difference: in Analytical Geometry all the results are inferred from the relations which exist between the quantities themselves, while in the Differential and Integral Calculus they are deduced by considering what may be indicated by *marks* drawn from variable quantities, under certain suppositions, and by *marks of such marks*.

Difference; In what it consists.

§ 285. Algebra, Analytical Geometry, the Differential and Integral Calculus, extended into the Theory of Variations, make up the subject of analytical science, of which Algebra is the elementary branch. As the limits of this work do not permit us to discuss the subject in full, we shall confine ourselves to Algebra, pointing out, occasionally, a few of the more obvious connections between it and the two other branches.

Analytical Science. Its parts. How far treated.

## ALGEBRA.

§ 286. On an analysis of the subject of Algebra, we think it will appear that the subject itself presents no serious difficulties, and that most of the embarrassment which is experienced by the pupil in gaining a knowledge of its principles, as well as in their applications, arises from not at-

Algebra. Difficulties. How overcome.

Language. tending sufficiently to the *language* or *signs* of the thoughts which are combined in the reasonings. At the hazard, therefore, of being a little diffuse, I shall begin with the very elements of the algebraic language, and explain, with much minuteness, the exact signification of the characters that stand for the quantities which are the subjects of the analysis; and also of those signs which indicate the several operations to be performed on the quantities.

Characters which represent quantity. Signs.

Quantities. How divided. § 287. The quantities which are the subjects of the algebraic analysis may be divided into two classes: those which are known or given, and those which are unknown or sought. The known are uniformly represented by the first letters of the alphabet, $a$, $b$, $c$, $d$, &c.; and the unknown by the final letters, $x$, $y$, $z$, $v$, $w$, &c.

How represented.

May be increased or diminished. § 288. Quantity is susceptible of being increased or diminished;* and there are five operations which can be performed upon a quantity that will give results differing from the quantity itself, viz.:

Five operations:

First. 1st. To add it to itself or to some other quantity;

* Section 75.

2d. To subtract some other quantity from it; Second.

3d. To multiply it by a number; Third.

4th. To divide it; Fourth.

5th. To extract a root of it. Fifth.

The cases in which the multiplier or divisor is 1, are of course excepted; as also the case where a root is to be extracted of 1. Exception.

§ 289. The five signs which denote these operations are too well known to be repeated here. These, with the signs of equality and inequality, the letters of the alphabet and the figures which are employed, make up the elements of the algebraic language. The words and phrases of the algebraic, like those of every other language, are to be taken in connection with each other, and are not to be interpreted as separate and isolated symbols.

Signs. Elements of the Algebraic language. Its words and phrases: How interpreted.

§ 290. The symbols of quantity are designed to represent quantity in general, whether abstract or concrete, whether known or unknown; and the signs which indicate the operations to be performed on the quantities are to be interpreted in a sense equally general. When the sign plus is written, it indicates that the quantity before which it is placed is to be added to some other quantity; and the sign minus implies the exist-

Symbols of quantity: General. Examples. Signs plus and minus.

ence of a minuend, from which the subtrahend is to be taken. One thing should be observed in regard to the signs which indicate the operations that are to be performed on quantities, viz. *they do not at all affect or change the nature of the quantity before or after which they are written, but merely indicate what is to be done with the quantity.* In Algebra, for example, the minus sign merely indicates that the quantity before which it is written is to be subtracted from some other quantity; and in Analytical Geometry, that the line before which it falls is estimated in a contrary direction from that in which it would have been reckoned, had it had the sign plus; but in neither case is the *nature* of the quantity itself different from what it would have been had it had the sign plus.

Signs have no effect on the nature of a quantity.

Examples: In Algebra.

In Analytical Geometry.

The interpretation of the language of Algebra is the first thing to which the attention of a pupil should be directed; and he should be drilled on the meaning and import of the symbols, until their significations and uses are as familiar as the sounds and combinations of the letters of the alphabet.

Interpretation of the language:

Its necessity.

§ 291. Beginning with the elements of the language, let any number or quantity be designated by the letter *a*, and let it be required to

Elements explained.

add this letter to itself, and find the result or sum. The addition will be expressed by

$$a + a = \text{the sum.}$$

But how is the sum to be expressed? By simply regarding $a$ as *one* $a$, or $1a$, and then observing that one $a$ and *one* $a$ make *two* $a$'s or $2\,a$: hence, Signification.

$$a + a = 2\,a;$$

and thus we place a figure before a letter to indicate how many times it is taken. Such figure is called a *Coefficient.* Coefficient.

§ 292. The product of several numbers is indicated by the sign of multiplication, or by simply writing the letters which represent the numbers by the side of each other. Thus, Product:

$$a \times b \times c \times d \times f, \text{ or } abcdf,$$

how indicated.

indicates the continued product of $a$, $b$, $c$, $d$, and $f$, and each letter is called a factor of the product: hence, a *factor* of a product is one of the multipliers which produce it. Any figure, as 5, written before a product, as Factor.

$$5\,abcdf,$$

is the coefficient of the product, and shows that the product is taken 5 times. Coefficient of a product.

18

Equal factors: what the product becomes.

§ 293. If the numbers represented by $a$, $b$, $c$, $d$, and $f$ were equal to each other, they would each be represented by a single letter $a$, and the product would then become

$$a \times a \times a \times a \times a = a^5;$$

How expressed.

that is, we indicate the product of several equal factors by simply writing the letter once and placing a figure above and a little at the right of it, to indicate how many times it is taken as a factor. The figure so written is called an *exponent*. Hence, an exponent is a simple form of expression, to point out how many equal factors are employed.

Exponent: where written.

Division: how expressed.

§ 294. The division of one quantity by another is indicated by simply writing the divisor below the dividend, after the manner of a fraction; by placing it on the right of the dividend with a horizontal line and two dots between them; or by placing it on the right with a vertical line between them: thus either form of expression:

Three forms.

$$\frac{b}{a}, \quad b \div a, \quad \text{or} \quad b \mid a$$

indicates the division of $b$ by $a$.

Roots: how indicated.

§ 295. The extraction of a root is indicated by the sign $\surd$. This sign, when used by itself indicates the lowest root, viz. the square root.

If any other root is to be extracted, as the third, fourth, fifth, &c., the figure marking the degree of the root is written above and at the left of the sign; as,

Index; where written.

$\sqrt[3]{}$ cube root, $\sqrt[4]{}$ fourth root, &c.

The figure so written, is called the *Index* of the root.

We have thus given the very simple and general language by which we indicate every one of the five operations that may be performed on an algebraic quantity, and *every process in Algebra* involves one or other of these operations.

Language for the five operations.

MINUS SIGN.

§ 296. The algebraic symbols are divided into two classes entirely distinct from each other, viz. the letters that are used to designate the quantities which are the subjects of the science, and the signs which are employed to indicate certain operations to be performed on those quantities. We have seen that all the algebraic processes are comprised under addition, subtraction, multiplication, division, and the extraction of roots; and it is plain, that the *nature* of a quantity is not at all changed by prefixing to it the sign which indicates either of these opera-

Algebraic language: how divided.

Algebraic processes: their number.

Do not change the nature of the quantities.

tions. The quantity denoted by the letter *a,* for example, is the same, in *every respect,* whatever sign may be prefixed to it; that is, whether it be added to another quantity, subtracted from it, whether multiplied or divided by any number, or whether we extract the square or cube or any other root of it. The algebraic signs, therefore, must be regarded merely as indicating *operations* to be performed on quantity, and not as affecting the *nature* of the quantities to which they may be prefixed. We say, indeed, that quantities are plus and minus, but this is an abbreviated language to express that they are to be added or subtracted.

Algebraic signs: how regarded.

Plus and Minus.

§ 297. In Algebra, as in Arithmetic and Geometry, all the principles of the science are deduced from the definitions and axioms; and the rules for performing the operations are but directions framed in conformity to such principles. Having, for example, fixed by definition, the power of the minus sign, viz. that any quantity before which it is written, shall be regarded as to be subtracted from another quantity, we wish to discover the process of performing that subtraction, so as to deduce therefrom a general *principle,* from which we can frame a rule applicable to all similar cases.

Principles of the science:

From what deduced.

Example.

What we wish to discover.

## SUBTRACTION.

§ 298. Let it be required, for example, to subtract from $b$ the difference between $a$ and $c$. Now, having written the letters, with their proper signs, the language of Algebra expresses that it is the *difference* only between $a$ and $c$, which is to be taken from $b$; and if this difference were known, we could make the subtraction at once. But the nature and generality of the algebraic symbols, enable us to *indicate operations,* merely, and we cannot in general make reductions until we come to the final result. In what general way, therefore, can we indicate the true difference?

Subtraction.

Process.

$$\begin{array}{l} b \\ a-c \\ \hline \end{array}$$

Difference.

Operations indicated.

If we indicate the subtraction of $a$ from $b$, we have $b-a$; but then we have taken away too much from $b$ by the number of units in $c$, for it was not $a$, but the *difference* between $a$ and $c$ that was to be subtracted from $b$. Having taken away *too much,* the remainder is *too small* by $c$: hence, if $c$ be added, the true remainder will be expressed by $b - a + c$.

Final formula.

$$\begin{array}{l} b-a \\ b-a+c \end{array}$$

Now, by analyzing this result, we see that the sign of every term of the subtrahend has been changed; and what has been shown with re-

Analysis of the result.

Generalization. spect to these quantities is equally true of all others standing in the same relation: hence, we have the following general rule for the subtraction of algebraic quantities:

Rule. *Change the sign of every term of the subtrahend, or conceive it to be changed, and then unite the quantities as in addition.*

MULTIPLICATION.

Multiplication. § 299. Let us now consider the case of multiplication, and let it be required to multiply $a - b$ by $c$. The algebraic language expresses that the difference between $a$ and $b$ is to be taken as many times as there are units in $c$. If we knew this difference, we could at once perform the multiplication. But by what general process is it to be performed without finding that difference? If we take $a$, $c$ times, the product will be $ac$; but as it was only the *difference* between $a$ and $b$, that was to be multiplied by $c$, this product $ac$ will be too great by $b$ taken $c$ times; that is, the true product will be expressed by $ac - bc$: hence, we see, that,

Signification of the language.

$$\begin{array}{r} a-b \\ c \\ \hline ac-bc \end{array}$$

Process:

Its nature.

Principle for the signs. *If a quantity having a plus sign be multiplied by another quantity having also a plus sign, the sign of the product will be plus; and if a quantity having a minus sign be multiplied by a quantity having a plus sign, the sign of the product will be minus.*

§ 300. Let us now take the most general case, viz. that in which it is required to multiply $a - b$ by $c - d$. General case:

Let us again observe that the algebraic language denotes that $a - b$ is to be taken as many times as there are units in $c - d$; and if these two differences were known, their product would at once form the product required. Its form.

$$\begin{array}{l} a-b \\ c-d \\ \hline ac-bc \\ \qquad\quad -ad+bd \\ \hline ac-bc-ad+bd \end{array}$$

First: let us take $a - b$ as many times as there are units in $c$; this product, from what has already been shown, is equal to $ac - bc$. But since the multiplier is not $c$, but $c - d$, it follows that this product is too large by $a - b$ taken $d$ times; that is, by $ad - bd$: hence, the first product diminished by this last, will give the true product. But, by the rule for subtraction, this difference is found by changing the signs of the subtrahend, and then uniting all the terms as in addition: hence, the true product is expressed by $ac - bc - ad + bd$. First step. Second step: How taken.

By analyzing this result, and employing an abbreviated language, we have the following gen- Analysis of the result.

eral principle to which the signs conform in multiplication, viz.:

General Principle.

*Plus multiplied by plus gives plus in the product; plus multiplied by minus gives minus; minus multiplied by plus gives minus; and minus multiplied by minus gives plus in the product.*

Remark.

§ 301. The remark is often made by pupils that the above reasoning appears very satisfactory so long as the quantities are presented under the above form; but why will $-b$ multiplied by $-d$ give plus $bd$? How can the product of two negative quantities *standing alone* be plus?

Particular case.

Minus sign:

In the first place, the minus sign being prefixed to $b$ and $d$, shows that in an *algebraic sense* they do not stand by themselves, but are connected with other quantities; and if they are not so connected, the minus sign makes no difference; for, it in no case affects the quantity, but merely points out a connection with other quantities. Besides, the product determined above, being independent of any particular value attributed to the letters $a$, $b$, $c$, and $d$, must be of such a form as to be true for all values; and hence for the case in which $a$ and $c$ are both equal to zero. Making this supposition, the product reduces to the form of $+bd$. The rules for the signs in division are readily deduced from

Its interpretation.

Form of the product: must be true for quantities of any value.

the definition of division, and the principles already laid down.

Signs in division.

ZERO AND INFINITY.

§ **302.** The terms zero and infinity have given rise to much discussion, and been regarded as presenting difficulties not easily removed. It may not be easy to frame a form of language that shall convey to a mind, but little versed in mathematical science, the precise ideas which these terms are designed to express; but we are unwilling to suppose that the ideas themselves are beyond the grasp of an ordinary intellect. The terms are used to designate the *two limits of Space and Number.*

Zero and Infinity.

Ideas not abstruse.

§ **303.** Assuming any two points in space, and joining them by a straight line, the distance between the points will be truly indicated by the length of this line, and this length may be expressed numerically by the number of times which the line contains a known unit. If now, the points are made to approach each other, the length of the line will diminish as the points come nearer and nearer together, until at length, when the two points become one, the length of the line will disappear, having attained its *limit,*

Illustration, showing the meaning of the term Zero.

which is called *zero*. If, on the contrary, the points recede from each other, the length of the line joining them will continually increase; but so long as the length of the line can be expressed in terms of a known unit of measure, it is not infinite. But, if we suppose the points removed, so that any known unit of measure would occupy no *appreciable portion* of the line, then the length of the line is said to be *Infinite*.

Illustration, showing the meaning of the term Infinity.

§ **304**. Assuming one as the unit of number, and admitting the self-evident truth that it may be increased or diminished, we shall have no difficulty in understanding the import of the terms zero and infinity, as applied to number. For, if we suppose the unit one to be continually diminished, by division or otherwise, the fractional units thus arising will be less and less, and in proportion as we continue the divisions, they will continue to diminish. Now, the limit or boundary to which these very small fractions approach, is called Zero, or nothing. So long as the fractional number forms an appreciable part of one, it is not zero, but a finite fraction; and the term zero is only applicable to that which forms no appreciable part of the standard.

The terms Zero and Infinity applied to numbers.

Illustration.

Zero:

Illustration. If, on the other hand, we suppose a number to be continually increased, the relation of this number to the unit will be constantly changing. So long as the number can be expressed in terms of the unit one, it is finite, and not infinite; but when the unit one forms no appreciable part of the number, the term infinite is used to express that state of value, or rather, that limit of value. Infinity;

§ 305. The terms zero and infinity are therefore employed to designate the limits to which decreasing and increasing quantities may be made to approach nearer than any assignable quantity; but these limits cannot be compared, in respect to magnitude, with any known standard, so as to give a finite ratio. The terms, how employed. Are limits.

§ 306. It may, perhaps, appear somewhat paradoxical, that zero and infinity should be defined as "the limits of number and space" when they are in themselves not measurable. But a limit is that "which sets bounds to, or circumscribes;" and as all finite space and finite number (and such only are implied by the terms Space and Number), are contained between zero and infinity, we employ these terms to designate the limits of Number and Space. Why limits? Definition of a limit. Of Space and Number.

## OF THE EQUATION.

Deductive reasoning. § 307. We have seen that all deductive reasoning involves certain processes of comparison, and that the syllogism is the formula to which those processes may be reduced.* It has also been stated that if two quantities be compared together, there will necessarily result the condition of equality or inequality. The equation is an analytical formula for expressing equality.

Comparison of quantities.

Condition.

Subject of equations: how divided. § 308. The subject of equations is divided into two parts. The first, consists in finding the equation; that is, in the process of expressing the relations existing between the quantities considered, by means of the algebraic symbols and formula. This is called the Statement of the proposition. The second is purely deductive, and consists, in Algebra, in what is called the solution of the equation, or finding the value of the unknown quantity; and in the other branches of analysis, it consists in the discussion of the equation; that is, in the drawing out from the equation every thing which it is capable of expressing.

First part: Statement.

Second part: Solution.

Discussion of an equation.

---

* Section 98.

§ 309. Making the statement, or finding the equation, is merely analyzing the problem, and expressing its elements and their relations in the language of analysis. It is, in truth, collating the facts, noting their bearing and connection, and inferring some general law or principle which leads to the formation of an equation.

Statement: what it is.

The condition of equality between two quantities is expressed by the sign of equality, which is placed between them. The quantity on the left of the sign of equality is called the first member, and that on the right, the second member of the equation. The first member corresponds to the subject of a proposition; the sign of equality is copula and part of the predicate, signifying, IS EQUAL TO. Hence, an equation is merely a proposition expressed algebraically, in which equality is predicated of one quantity as compared with another. It is the great formula of analysis.

Equality of two quantities: How expressed. 1st member. 2d member. Subject. Predicate. Proposition.

§ 310. We have seen that every quantity is either abstract or concrete:* hence, an equation, which is a general formula for expressing equality, must be either abstract or concrete.

Abstract. Concrete.

An abstract equation expresses merely the

* Section 75.

relation of equality between two abstract quantities: thus,

Abstract equation.

$$a + b = x,$$

is an abstract equation, if no unit of value be assigned to either member; for, until that be done the abstract unit one is understood, and the formula merely expresses that the sum of $a$ and $b$ is equal to $x$, and is true, equally, of all quantities.

Concrete equation.

But if we assign a concrete unit of value, that is, say that $a$ and $b$ shall each denote so many pounds weight, or so many feet or yards of length, $x$ will be of the same denomination, and the equation will become concrete or denominate.

Five operations may be performed.

§ 311. We have seen that there are five operations which may be performed on an algebraic quantity.* We assume, as an axiom, that if the same operation, under either of these processes, be performed on both members of an equation, the equality of the members will not be changed. Hence, we have the five following

Axioms.

AXIOMS.

First.

1. If equal quantities be added to both members of an equation, the equality of the members will not be destroyed.

---

* Section 288.

2. If equal quantities be subtracted from both members of an equation, the equality will not be destroyed. Second.

3. If both members of an equation be multiplied by the same number, the equality will not be destroyed. Third.

4. If both members of an equation be divided by the same number, the equality will not be destroyed. Fourth.

5. If the same root of both members of an equation be extracted, the equality of the members will not be destroyed. Fifth.

Every operation performed on an equation will fall under one or other of these axioms, and they afford the means of solving all equations which admit of solution. Use of axioms.

§ 312. The term Equality, in Geometry, expresses that relation between two magnitudes which will cause them to coincide, throughout their whole extent, when applied to each other. Equality: Its meaning in Geometry. The same term, in Algebra, merely implies that the quantity, of which equality is predicated, and that to which it is affirmed to be equal, contain the same unit of measure an equal number of times: hence, the algebraic signification of the term equality corresponds to the signification of the geometrical term equivalency. Its meaning in Algebra. Corresponds to equivalency.

§ 313. We have thus pointed out some of the marked characteristics of analysis. In Algebra, the elementary branch, the quantities, about which the science is conversant, are divided, as has been already remarked, into known and unknown, and the connections between them, expressed by the equation, afford the means of tracing out further relations, and of finding the values of the unknown quantities in terms of the known.

Classes of quantities in Algebra.

In the other branches of analysis, the quantities considered are divided into two general classes, Constant and Variable; the former preserving fixed values throughout the same process of investigation, while the latter undergo changes of value according to fixed laws; and from such changes we deduce, by means of the equation, common principles, and general properties applicable to all quantities.

How divided in the other branches of Analysis.

§ 314. The correspondence between the processes of reasoning, as exhibited in the subject of general logic, and those which are employed in mathematical science, is readily accounted for, when we reflect, that the reasoning process is essentially the same in all cases; and that any change in the language employed, or in the subject to which the reasoning is applied, does not

Correspondence in methods of reasoning accounted for.

at all change the nature of the process, or materially vary its form.

§ 315. We shall not pursue the subject of analysis any further; for, it would be foreign to the purposes of the present work to attempt more than to point out the general features and characteristics of the different branches of mathematical science, to present the subjects about which the science is conversant, to explain the peculiarities of the language, the nature of the reasoning processes employed, and of the connecting links of that golden chain which binds together all the parts, forming an harmonious whole.

Objects of the present work:

How far extended.

### SUGGESTIONS FOR THOSE WHO TEACH ALGEBRA.

1. Be careful to explain that the letters employed, are the mere symbols of quantity. That of, and in themselves, they have no meaning or signification whatever, but are used merely as the signs or representatives of such quantities as they may be employed to denote.

Letters are but mere symbols.

2. Be careful to explain that the signs which are used are employed merely for the purpose of indicating the five operations which may be performed on quantity; and that they indicate

Signs indicate operations.

operations merely, without at all affecting the nature of the quantities before which they are placed.

Letters and signs elements of language.

3. Explain that the letters and signs are the elements of the algebraic language, and that the language itself arises from the combination of these elements.

Algebraic formula:

4. Explain that the finding of an algebraic formula is but the translation of certain ideas, first expressed in our common language, into the language of Algebra; and that the interpretation of an algebraic formula is merely translating its various significations into common language.

Its interpretation.

Language.

5. Let the language of Algebra be carefully studied, so that its construction and significations may be clearly apprehended.

Coefficient, Exponent.

6. Let the difference between a coefficient and an exponent be carefully noted, and the office of each often explained; and illustrate frequently the signification of the language by attributing numerical values to letters in various algebraic expressions.

Similar quantities.

7. Point out often the characteristics of similar and dissimilar quantities, and explain which may be incorporated and which cannot.

Minus sign.

8. Explain the power of the minus sign, as shown in the four ground rules, but very par-

ticularly as it is illustrated in subtraction and multiplication.

9. Point out and illustrate the correspondence between the four ground rules of Arithmetic and Algebra; and impress the fact, that their differences, wherever they appear, arise merely from differences in notation and language: the principles which govern the operations being the same in both. Arithmetic and Algebra compared.

10. Explain with great minuteness and particularity all the characteristic properties of the equation; the manner of forming it; the different kinds of quantity which enter into its composition; its examination or discussion; and the different methods of elimination. Equation. Its properties.

11. In the equation of the second degree, be careful to dwell on the four forms which embrace all the cases, and illustrate by many examples that every equation of the second degree may be reduced to one or other of them. Explain very particularly the meaning of the term root; and then show, why every equation of the first degree has one, and every equation of the second degree two. Dwell on the properties of these roots in the equation of the second degree. Show why their sum, in all the forms, is equal to the coefficient of the second term, taken with a contrary sign; and why their Equation of the second degree. Its forms. Its roots. Their sum.

Their product. product is equal to the absolute term with a contrary sign. Explain when and why the roots are imaginary.

General Principles: 12. In fine, remember that every operation and rule is based on a principle of science, and that an intelligible reason may be given for it. Find that reason, and impress it on the mind of your pupil in plain and simple language, and by familiar and appropriate illustrations. You will thus impress right habits of investigation and study, and he will grow in knowledge. The broad field of analytical investigation will be opened to his intellectual vision, and he will have made the first steps in that sublime science which discovers the laws of nature in their most secret hiding-places, and follows them, as they reach out, in omnipotent power, to control the motions of matter through the entire regions of occupied space.

Should be explained.

They lead to general laws.

# BOOK III.

# UTILITY OF MATHEMATICS.

## CHAPTER I.

### THE UTILITY OF MATHEMATICS CONSIDERED AS A MEANS OF INTELLECTUAL TRAINING AND CULTURE.

First efforts.

§ 316. The first efforts in mathematical science are made by the child in the process of counting. He counts his fingers, and repeats the words one, two, three, four, five, six, seven, eight, nine, ten, until he associates with these words the ideas of one or more, and thus acquires his first notions of number. Hence, the idea of number is first presented to the mind by means of sensible objects; but when once clearly apprehended, the perception of the sensible objects fades away, and the mind retains only the abstract idea. Thus, the child, after counting for a time with the aid of his fingers or his marbles, dispenses with these cumbrous helps, and

Counting of sensible objects.

Generalization.

Abstraction.

employs only the abstract ideas, which his mind embraces with clearness and uses with facility.

Analytical method: Uses sensible objects at first.

§ 317. In the first stages of the analytical methods, where the quantities considered are represented by the letters of the alphabet, sensible objects again lend their aid to enable the mind to gain exact and distinct ideas of the things considered; but no sooner are these ideas obtained than the mind loses sight of the things themselves, and operates entirely through the instrumentality of symbols.

Geometry. First impressions by sensible objects.

§ 318. So, also, in Geometry. The right line may first be presented to the mind, as a black mark on paper, or a chalk mark on a blackboard, to impress the geometrical definition, that "A straight line does not change its direction between any two of its points." When this definition is clearly apprehended, the mind needs no further aid from the eye, for the image is forever imprinted.

A plane. Definition: How illustrated.

§ 319. The idea of a plane surface may be impressed by exhibiting the surface of a polished mirror; and thus the mind may be aided in apprehending the definition, that "a plane surface is one in which, if any two points be taken, the straight line which joins them will lie wholly in the surface." But when the definition is understood, the mind requires no sensible object to aid its conception. The ideal alone fills the mind, and the image lives there without any connection with sensible objects.

Its true conception.

§ 320. Space is indefinite extension, in which all bodies are situated. A solid or body is any portion of space embracing the three dimensions of length, breadth, and thickness. To give to the mind the true conception of a solid, the aid of the eye may at first be necessary; but the idea being once impressed, that a solid, in a strictly mathematical sense, means only a portion of space, and has no reference to the matter with which the space may be filled, the mind turns away from the material object, and dwells alone on the ideal.

Space. Solid: How conceived. What it really is.

§ 321. Although quantity, in its general sense, is the subject of mathematical inquiry, yet the language of mathematics is so constructed, that the investigations are pursued without the slightest reference to quantity as a material substance. We have seen that a system of symbols, by which quantities may be represented, has been adopted, forming a language for the expression

Quantity: Language: How constructed. Symbols:

of ideas entirely disconnected from material objects, and yet capable of expressing and representing such objects. This symbolical language, at once copious and exact, not only enables us to express our known thoughts, in every department of mathematical science, but is a potent means of pushing our inquiries into unexplored regions, and conducting the mind with certainty to new and valuable truths.

Nature of the language: What it accomplishes.

Advantages of an exact language.

§ 322. The nature of that culture, which the mind undergoes by being trained in the use of an exact language, in which the connection between the *sign* and the *thing signified* is unmistakable, has been well set forth by a living author, greatly distinguished for his scientific attainments.* Of the pure sciences, he says:

Herschel's views.

"Their objects are so definite, and our notions of them so distinct, that we can reason about them with an assurance that the words and signs of our reasonings are full and true representatives of the things signified; and, consequently, that when we use language or signs in argument, we neither by their use introduce extraneous notions, nor exclude any part of the case before us from consideration. For exam-

Exact language prevents error.

* Sir John Herschel, Discourse on the study of Natural Philosophy.

ple: the words space, square, circle, a hundred, &c., convey to the mind notions so complete in themselves, and so distinct from every thing else, that we are sure when we use them we know and have in view the whole of our own meaning. It is widely different with words expressing natural objects and mixed relations. Take, for instance, *Iron*. Different persons attach very different ideas to this word. One who has never heard of magnetism has a widely different notion of iron from one in the contrary predicament. The vulgar who regard this metal as incombustible, and the chemist, who sees it burn with the utmost fury, and who has other reasons for regarding it as one of the most combustible bodies in nature; the poet, who uses it as an emblem of rigidity; and the smith and engineer, in whose hands it is plastic, and moulded like wax into every form; the jailer, who prizes it as an obstruction, and the electrician, who sees in it only a channel of open communication by which that most impassable of obstacles, the air, may be traversed by his imprisoned fluid,—have all different, and all imperfect notions of the same word. The meaning of such a term is like the rainbow—everybody sees a different one, and all maintain it to be the same."

Mathematical terms exact.

Different in regard to other terms.

How iron is regarded by the chemist:

The poet:

The jailer:

The electrician.

Final illustration.

"It is, in fact, in this double or incomplete

Incomplete meaning the source of error. sense of words, that we must look for the origin of a very large portion of the errors into which we fall. Now, the study of the abstract sciences, Mathematics free from such errors. such as Arithmetic, Geometry, Algebra, &c., while they afford scope for the exercise of reasoning about objects that are, or, at least, may be conceived to be, external to us; yet, being free from these sources of error and mistake, Requires a strict use of language. accustom us to the strict use of language as an instrument of reason, and by familiarizing us in our progress towards truth, to walk uprightly and straightforward, on firm ground, give us that proper and dignified carriage of mind which Results. could never be acquired by having always to pick our steps among obstructions and loose fragments, or to steady them in the reeling tempests of conflicting meanings."

Two ways of acquiring knowledge. § 323. Mr. Locke lays down two ways of increasing our knowledge:

1st. "Clear and distinct ideas with settled names; and,

2d. "The finding of those which show their agreement or disagreement;" that is, the searching out of new ideas which result from the combination of those that are known.

First. In regard to the first of these ways, Mr. Locke says: "The first is to get and settle in our minds determined ideas of those things, whereof we have general or specific names; at least, of so many of them as we would consider and improve our knowledge in, or reason about." * * * "For, it being evident, that our knowledge cannot exceed our ideas, as far as they are either imperfect, confused, or obscure, we cannot expect to have certain, perfect, or clear knowledge."

Ideas of things must be distinct.

Reason.

§ 324. Now, the ideas which make up our knowledge of mathematical science, fulfil exactly these requirements. They are all impressed on the mind by a fixed, definite, and certain language, and the mind embraces them as so many images or pictures, clear and distinct in their outlines, with names which suggest at once their characteristics and properties.

Why it is so in mathematics.

§ 325. In the second method of increasing our knowledge, pointed out by Mr. Locke, mathematical science offers the most ample and the surest means. The reasonings are all based on self-evident truths, and are conducted by means of the most striking relations between the known and the unknown. The things reasoned about, and the methods of reasoning, are so clearly apprehended, that the mind never hesitates or doubts. It comprehends, or it does not compre-

Second.

Why mathematics offer the surest means.

hend, and the line which separates the known from the unknown, is always well defined. These characteristics give to this system of reasoning a superiority over every other, arising, not from any difference in the logic, but from a difference in the things to which the logic is applied. Observation may deceive, experiment may fail, and experience prove treacherous, but demonstration never.

Characteristics of the reasoning. Its advantages. Demonstration certain.

"If it be true, then, that mathematics include a perfect system of reasoning, whose premises are self-evident, and whose conclusions are irresistible, can there be any branch of science or knowledge better adapted to the improvement of the understanding? It is in this capacity, as a strong and natural adjunct and instrument of reason, that this science becomes the fit subject of education with all conditions of society, whatever may be their ultimate pursuits. Most sciences, as, indeed, most branches of knowledge, address themselves to some particular taste, or subsequent avocation; but this, while it is before all, as a useful attainment, especially adapts itself to the cultivation and improvement of the thinking faculty, and is alike necessary to all who would be governed by reason, or live for usefulness."*

Mathematics includes a certain system. An adjunct and instrument of reason; and necessary to all.

* Mansfield's Discourse on the Mathematics.

§ 326. The following, among other considerations, may serve to point out and illustrate the value of mathematical studies, as a means of mental improvement and development.

Considerations of the value of mathematics.

1. We readily conceive and clearly apprehend the things of which the science treats; they being things simple in themselves and readily presented to the mind by plain and familiar language. For example: the idea of number, of one or more, is among the first ideas implanted in the mind; and the child who counts his fingers or his marbles, understands the art of numbering them as perfectly as he can know any thing. So, likewise, when he learns the definition of a straight line, of a triangle, of a square, of a circle, or of a parallelogram, he conceives the idea of each perfectly, and the name and the image are inseparably connected. These ideas, so distinct and satisfactory, are expressed in the simplest and fewest terms, and may, if necessary, be illustrated by the aid of sensible objects.

First. They give clear conceptions of things.

Example.

They establish clear relations between definitions and things.

2. The words employed in the definitions are always used in the same sense—each expressing at all times the same idea; so that when a definition is apprehended, the conception of the thing, whose name is defined, is perfect in the mind.

Second. Words are always used in the same sense.

There is, therefore, no doubt or ambiguity

Hence, it is certain.

either in the language, or in regard to what is affirmed or denied of the things spoken of; but all is certainty, both in the language employed and in the ideas which it expresses.

Third. It employs no definition or axiom not evident and clear.

3. The science of mathematics employs no definition which may not be clearly comprehended—lays down no axioms not universally true, and to which the mind, by the very laws of its nature, readily assents; and because, also, in the process of the reasoning, no principle or truth is taken for granted, but every link in the chain of the argument is immediately connected with a definition or axiom, or with some principle previously established.

The connection evident.

Fourth. The order strengthens different faculties.

4. The order established in presenting the subject to the mind, aids the memory at the same time that it strengthens and improves the reasoning powers. For example: first, there are the definitions of the names of the things which are the subjects of the reasoning; then the axioms, or self-evident truths, which, together with the definitions, form the basis of the science. From these the simplest propositions are deduced, and then follow others of greater difficulty; the whole connected together by rigorous logic—each part receiving strength and light from all the others. Whence, it follows, that any proposition may be traced to first prin-

How ideas are presented.

How the deductions follow.

ciples; its dependence upon and connection with those principles made obvious; and its truth established by certain and infallible argument. Propositions traced to their sources.

5. The demonstrative argument of mathematics produces the most certain knowledge of which the mind is susceptible. It establishes truth so clearly, that none can doubt or deny. For, if the premises are certain—that is, such that all minds admit their truth without hesitation or doubt, and if the method of drawing the conclusions be lawful—that is, in accordance with the infallible rules of logic, the inferences must also be true. Truths thus established may be relied on for their verity; and the knowledge thus gained may well be denominated SCIENCE.

Fifth. Argument the most certain.

Reasons.

Such knowledge science.

§ 327. There are, as we have seen, in mathematics, two systems of investigation quite different from each other: the Synthetical and the Analytical; the synthetical beginning with the definitions and axioms, and terminating in the highest truth reached by Geometry.

Two systems: Synthesis, Analysis.

"This science presents the very method by which the human mind, in its progress from childhood to age, develops its faculties. What first meets the observation of a child? Upon what are his earliest investigations employed?

Synthetical.

First notions.

What is first observed.

Next to color, which exists only to the sight, figure, extension, dimension, are the first objects which he meets, and the first which he examines. He ascertains and acknowledges their existence; then he perceives plurality, and begins to enumerate; finally he begins to draw conclusions from the parts to the whole, and makes a law from the individual to the species. Thus, he has obtained figure, extension, dimension, enumeration, and generalization. This is the teaching of nature; and hence, when this process becomes embodied in a perfect system, as it is in Geometry, that system becomes the easiest and most natural means of strengthening the mind in its early progress through the fields of knowledge."

Progress of inquiry.

Process developed in the system of Geometry.

First necessity for Analysis:

"Long after the child has thus begun to generalize and deduce laws, he notices objects and events, whose exterior relations afford no conclusion upon the subject of his contemplation. Machinery is in motion—effects are produced. He is surprised; examines and inquires. He reasons backward from effect to cause. This is *Analysis,* the metaphysics of mathematics; and through all its varieties—in Arithmetic—in Algebra—and in the Differential and Integral Calculus, it furnishes a grand armory of weapons for acute philosophical investigation. But analysis

Its method.

What the science is:

advances one step further by its peculiar notation; it exercises, in the highest degree, the faculty of abstraction, which, whether morally or intellectually considered, is always connected with the loftiest efforts of the mind. Thus this science comes in to assist the faculties in their progress to the ultimate stages of reasoning; and the more these analytical processes are cultivated, the more the mind looks in upon itself, estimates justly and directs rightly those vast powers which are to buoy it up in an eternity of future being."*

What it does: What it finally accomplishes.

§ 328. To the quotations, which have already been so ample, we will add but two more.

"In the mathematics, I can report no deficience, except it be that men do not sufficiently understand the excellent use of the pure mathematics, in that they do remedy and cure many defects in the wit and faculties intellectual. For, if the wit be too dull, they sharpen it; if too wandering, they fix it; if too inherent in the sense, they abstract it."† Again:

Bacon's opinion of mathematics.

"Mathematics serve to inure and corroborate the mind to a constant diligence in study, to

How the study of

* Mansfield's Discourses on Mathematics.

† Lord Bacon.

mathematics affects the mind. undergo the trouble of an attentive meditation, and cheerfully contend with such difficulties as lie in the way. They wholly deliver us from credulous simplicity, most strongly fortify us against the vanity of skepticism, effectually restrain us from a rash presumption, most easily incline us to due assent, perfectly subjugate us to the government and weight of reason, and inspire us with resolution to wrestle against the injurious tyranny of false prejudices.

Its influences.

How they are exerted.

"If the fancy be unstable and fluctuating, it is, as it were, poised by this ballast, and steadied by this anchor; if the wit be blunt, it is sharpened by this whetstone; if it be luxuriant, it is pruned by this knife; if it be headstrong, it is restrained by this bridle; and if it be dull, it is roused by this spur."*

Mathematics a pure science.

§ 329. Mathematics, in all its branches, is, in fact, a science of ideas alone, unmixed with matter or material things; and hence, is properly termed a Pure Science. It is, indeed, a fairy land of the pure ideal, through which the mind is conducted by conventional symbols, as thought is conveyed along wires constructed by the hand of man.

* Dr. Barrow.

§ 330. In conclusion, therefore, we may claim for the study of Mathematics, that it impresses the mind with clear and distinct ideas; cultivates habits of close and accurate discrimination; gives, in an eminent degree, the power of abstraction; sharpens and strengthens all the faculties, and develops, to their highest range, the reasoning powers. The tendency of this study is to raise the mind from the servile habit of imitation to the dignity of self-reliance and self-action. It arms it with the inherent energies of its own elastic nature, and urges it out on the great ocean of thought, to make new discoveries, and enlarge the boundaries of mental effort.

What may fairly be claimed for mathematics.

Its tendency.

The reasons.

## CHAPTER II.

### THE UTILITY OF MATHEMATICS REGARDED AS A MEANS OF ACQUIRING KNOWLEDGE—BACONIAN PHILOSOPHY.

Mathematics: § **331**. In the preceding chapter, we considered the effects of mathematical studies on the mind, merely as a means of discipline and training. How considered heretofore: We regarded the study in a single point of view, viz. as the drill-master of the intellectual faculties — the power best adapted to bring them all into order—to impart strength, and to give to them organization. In the How now considered. present chapter we shall consider the study under a more enlarged aspect—as furnishing to man the keys of hidden and precious knowledge, and as opening to his mind the whole volume of nature.

Material Universe. § **332**. The material universe, which is spread out before us, is the first object of our rational regards. Material things are the first with which we have to do. The child plays with his toys in the nursery, paddles in the limpid water, twirls his top, and strikes with the hammer. At a maturer age a higher class of ideas are embraced. The earth is surveyed, teeming with its products, and filled with life. Man looks around him with wondering and delighted eyes. The earth he stands upon appears to be made of firm soil and liquid waters. The land is broken into an irregular surface by abrupt hills and frowning mountains. The rivers pursue their courses through the valleys, without any apparent cause, and finally seem to lose themselves in their own expansion. He notes the return of day and night, at regular intervals, turns his eyes to the starry heavens, and inquires how far those sentinels of the night may be from the world they look down upon. He is yet to learn that all is governed by general laws imparted by the fiat of Him who created all things; that matter, in all its forms, is subject to those laws; and that man possesses the capacity to investigate, develop, and understand them. It is of the essence of law that it includes all possible contingencies, and insures implicit obedience; and such are the laws of nature.

Elements of knowledge:

Obtained by observation.

Course of nature:

Governed by general laws:

Man possesses the faculty to investigate and understand them.

§ 333. To the man of chance, nothing is more mysterious than the developments of science.

Uniformity: He does not see how so great a uniformity can

Variety: consist with the infinite variety which pervades every department of nature. While no two individuals of a species are exactly alike, the resemblance and conformity are so close, that the naturalist, from the examination of a single bone, finds no difficulty in determining the species, size, and structure of the animal. So,

They appear in all things. also, in the vegetable and mineral kingdoms: all the structures of growth or formation, although infinitely varied, are yet conformable to like general laws.

Science necessary to the development of law. This wonderful mechanism, displayed in the structure of animals, was but imperfectly understood, until touched by the magic wand of science. Then, a general law was found to pervade the whole. Every bone is of that length

What science shows. and diameter best adapted to its use; every muscle is inserted at the right point, and works about the right centre; the feathers of every bird are shaped in the right form, and the curves in which they cleave the air are best adapted

What may be demonstrated. to velocity. It is demonstrable, that in every case, and in all the variety of forms in which forces are applied, either to increase power or gain velocity, the very best means have been adopted to produce the desired result. And why should it not be so, since they are employed by the all-wise Architect? Why it is so.

§ 334. It is in the investigations of the laws of nature that mathematics finds its widest range and its most striking applications. Applications of Mathematics.

Experience, aided by observation and enlightened by experiment, is the recognised fountain of all knowledge of nature. On this foundation Bacon rested his Philosophy. He saw that the Deductive process of Aristotle, in which the conclusions do not reach beyond the premises, was not progressive. It might, indeed, improve the reasoning powers, cultivate habits of nice discrimination, and give great proficiency in verbal dialectics; but the basis was too narrow for that expansive philosophy, which was to unfold and harmonize all the laws of nature. Hence, he suggested a careful examination of nature in every department, and laid the foundations of a new philosophy. Nature was to be interrogated by experiment, observation was to note the results, and gather the facts into the storehouse of knowledge. Facts, so obtained, were subjected to analysis and collation, and general laws inferred from such classification by

Bacon's Philosophy.

Aristotle's:

Its defects.

What Bacon suggested.

The means to be employed.

Bacon's system Inductive. a reasoning process called Induction. Hence, the system of Bacon is said to be Inductive.

New Philosophy: § 335. This new philosophy gave a startling impulse to the human mind. Its subject was Nature—material and immaterial; its object, the discovery and analysis of those general laws What it did. which pervade, regulate, and impart uniformity to all things; its processes, experience, experiment, and observation for the ascertainment of Its nature. facts; analysis and comparison for their classification; and reasoning, for the establishment of What aided it. general laws. But the work would have been incomplete without the aid of deductive science. General laws deduced from many separate cases, What it needed. by Induction, needed additional proof; for, they might have been inferred from resemblances too slight, or coincidences too few. Mathematical science affords such proofs.

The truths of Induction: § 336. Regarding general laws, established by Induction, as fundamental truths, expressing these by means of the analytical formulas, and then operating on these formulas by the known pro- How verified by Analysis. cesses of mathematical science, we are enabled, not only to verify the truths of induction, but often to establish new truths, which were hidden from experiment and observation. As the inductive process may involve error, while the deductive cannot, there are weighty scientific reasons, for giving to every science as much of the character of a Deductive Science as possible. Every science, therefore, should be constructed with the fewest and simplest possible inductions. These should be made the basis of deductive processes, by which every truth, however complex, should be proved, even if we chose to verify the same by induction, based on specific experiments.

As far as possible, all sciences should be made Deductive.

§ 337. Every branch of Natural Philosophy was originally experimental; each generalization rested on a special induction, and was derived from its own distinct set of observations and experiments. From being sciences of pure experiment, as the phrase is, or, to speak more correctly, sciences in which the reasonings consist of no more than one step, and that a step of induction; all these sciences have become, to some extent, and some of them in nearly their whole extent, sciences of pure reasoning: thus, multitudes of truths, already known by induction, from as many different sets of experiments, have come to be exhibited as deductions, or corollaries from inductive propositions of a simpler and more universal character. Thus, mechan-

Natural Philosophy was experimental:

Is now deductive.

Mathematical or

Deductive Sciences: ics, hydrostatics, optics, and acoustics, have successively been rendered mathematical; and astronomy was brought by Newton within the laws of general mechanics.

Their advantages: The substitution of this circuitous mode of proceeding for a process apparently much easier and more natural, is held, and justly too, to be the greatest triumph in the investigation of nature. They rest on Inductions. But, it is necessary to remark, that although, by this progressive transformation, all sciences tend to become more and more deductive, they are not, therefore, the less inductive; for, every step in the deduction rests upon an antecedent induction. Sciences deductive or experimental. The opposition is, perhaps, not so much between the terms Deductive and Inductive as between Deductive and Experimental.

Experimental Science: § 338. A science is experimental, in proportion as every new case, which presents any peculiar features, stands in need of a new set of observations and experiments, and a fresh induction. It is deductive, in proportion as it can When deductive. draw conclusions, respecting cases of a new kind, by processes which bring these cases under old inductions, or show them to possess known marks of certain attributes.

§ 339. We can now, therefore, perceive, what

is the generic distinction between sciences that can be made deductive and those which must, as yet, remain experimental. The difference consists in our having been able, or not yet able, to draw from first inductions as from a general law, a series of connected and dependent truths. When this can be done, the deductive process can be applied, and the science becomes deductive. For example: when Newton, by observing and comparing the motions of several of the heavenly bodies, discovered that all the motions, whether regular or apparently anomalous, of all the observed bodies of the Solar System, conformed to the law of moving around a common centre, urged by a centripetal force, varying directly as the mass, and inversely as the square of the distance from the centre, he *inferred the existence of such a law for all the bodies of the system,* and then demonstrated, by the aid of mathematics, that no other law could produce the motions. This is the greatest example which has yet occurred of the transformation, at one stroke, of a science which was in a great degree purely experimental, into a deductive science.

Difference between Deductive and Experimental Sciences.

Deductive.

Example.

What Newton inferred:

What he proved.

§ **340.** How far the study of mathematics prepares the mind for such contemplations and

Study of mathematics:

prepares the mind.

such knowledge, is well set forth by an old writer, himself a distinguished mathematician. He says:

Dr. Barrow's opinion.

"The steps are guided by no lamp more clearly through the dark mazes of nature, by no thread more surely through the infinite turnings of the labyrinth of philosophy; nor lastly, is the bottom of truth sounded more happily by any other line. I will not mention with how plentiful a stock of knowledge the mind is furnished from these; with what wholesome food it is nourished, and what sincere pleasure it enjoys. But if I speak further, I shall neither be the only person nor the first, who affirms it, that while the mind is abstracted, and elevated from sensible matter, distinctly views pure forms, conceives the beauty of ideas, and investigates the harmony of proportions, the manners themselves are sensibly corrected and improved, the affections composed and rectified, the fancy calmed and settled, and the understanding raised and excited to more divine contemplations: all of which I might defend by the authority and confirm by the suffrages of the greatest philosophers."*

How mathematics furnish the mind:

Abstract and elevate it:

Confirmed by philosophers.

Herschel's

§ 341. Sir John Herschel, in his Introduction

* Dr. Barrow.

to his admirable Treatise on Astronomy, very justly remarks, that,

opinions.

"Admission to its sanctuary [the science of Astronomy], and to the privileges and feelings of a votary, is only to be gained by one means—*sound and sufficient knowledge of mathematics, the great instrument of all exact inquiry, without which no man can ever make such advances in this or any other of the higher departments of science as can entitle him to form an independent opinion on any subject of discussion within their range.*

Mathematical science, indispensable to a knowledge of Astronomy.

"It is not without an effort that those who possess this knowledge can communicate on such subjects with those who do not, and adapt their language and their illustrations to the necessities of such an intercourse. Propositions which to the one are almost identical, are theorems of import and difficulty to the other; nor is their evidence presented in the same way to the mind of each. In treating such propositions, under such circumstances, the appeal has to be made, not to the pure and abstract reason, but to the sense of analogy—to practice and experience: principles and modes of action have to be established, not by direct argument from acknowledged axioms, but by continually referring to the sources from which the axioms them-

Information cannot be given to such as have no mathematics:

Except by very cumbrous methods.

Reasons:

Must begin with the simplest elements:

selves have been drawn, viz. examples; that is to say, by bringing forward and dwelling on simple and familiar instances in which the same principles and the same or similar modes of action take place; thus erecting, as it were, in each particular case, a separate induction, and constructing at each step a little body of science to meet its exigencies. The difference is that of pioneering a road through an untraversed country, and advancing at ease along a broad and beaten highway; that is to say, if we are determined to make ourselves distinctly understood, and will appeal to reason at all." Again:

Illustration of the difference between instruction by scientific and unscientific methods.

Mathematics necessary to physics:

"A certain moderate degree of acquaintance with abstract science is highly desirable to every one who would make any considerable progress in physics. As the universe exists in time and place; and as motion, velocity, quantity, number, and order, are main elements of our knowledge of external things and their changes, an acquaintance with these, abstractedly considered (that is to say, independent of any consideration of particular things moved, measured, counted, or arranged), must evidently be a useful preparation for the more complex study of nature."*

Why it is so necessary.

* Sir John Herschel on the study of Natural Philosophy.

§ 342. If we consider the department of chemistry,—which analyzes matter, examines the elements of which it is composed, develops the laws which unite these elements, and also the agencies which will separate and reunite them,—we shall find that no intelligent and philosophical analysis can be made without the aid of mathematics. Necessary in chemistry.

§ 343. The mechanism of the physical universe, and the laws which govern and regulate its motions, were long unknown. As late as the 17th century, Galileo was imprisoned for promulgating the theory that the earth revolves on its axis; and to escape the fury of persecution, renounced the deductions of science. Now, every student of a college, and every ambitious boy of the academy, may, by the aid of his Algebra and Geometry, demonstrate the existence and operation of those general laws which enable him to trace with certainty the path and motions of every body which circles the heavens. Laws long unknown. Galileo. His theory. Now known to all: By what means demonstrated.

§ 344. What knowledge is more precious, or more elevating to the mind, than that which assures us that the solar system, of which the sun is the centre, and our earth one of the smaller bodies, is governed by the general law of gravitation; that is, that each body is retained in its orbit by attracting, and being at- Value of scientific knowledge:

What it teaches.

tracted by, all the others? This power of attraction, by which matter operates on matter, is the great governing principle of the material world. The motion of each body in the heavens depends on the forces of attraction of all the others; hence, to estimate such forces—varying as they do with the quantity of matter in each body, and inversely as the squares of their distances apart—is no easy problem; yet analysis has solved it, and with such certainty, that the exact spot in the heavens may be marked at which each body will appear at the expiration of any definite period of time. Indeed, a telescope may be so arranged, that at the end of that time either one of the heavenly bodies would present itself to the field of view; and if the instrument could remain fixed, though the time were a thousand years, the precise moment would discover the planet to the eye of the observer, and thus attest the certainty of science.

The things not easy.

Analysis:

What it has done:

How a result might be verified by experiment.

§ 345. But analysis has done yet more. It has not only measured the attractive power of each of the heavenly bodies; determined their distances from a common point and from each other; ascertained their specific gravities and traced their orbits through the heavens; but has also discovered the existence of balancing

Analysis determines balancing forces.

and conservative forces, evincing the highest evidence of contrivance and design.

Evidence of design.

§ 346. A superficial view of the architecture of the heavens might inspire a doubt of the stability of the entire system. The mutual action of the bodies on each other produces what is called an irregularity in their motions. The earth, for example, in her annual course around the sun, is affected by the attraction of the moon and of all the planets which compose the solar system; and these attracting forces appear to give an irregularity to her motions. The moon in her revolutions around the earth is also influenced by the attraction of the sun, the earth, and of all the other planets, and yields to each a motion exactly proportionate to the force exerted; and the same is equally true of all the bodies which belong to the system. It was reserved for analysis to demonstrate that every supposed irregularity of motion is but the consequence of a general law; that every change is constancy, and every diversity uniformity. Thus, mathematical science assures us that our system has not been abandoned to blind chance, but that a superintending Providence is ever exerted through those general laws, which are so minute as to govern the motions of the feather as it is

Architecture of the heavens shows permanency.

Example of the earth:

Of the moon.

Of the other planets.

Mathematics proves the permanency of the system.

Generality of laws. wafted along on the passing breeze, and yet so omnipotent as to preserve the stability of worlds.

§ 347. But analysis goes yet another step. That class of wandering bodies, known to us by the name of comets, although apparently escaped from their own spheres, and straying heedlessly through illimitable space, have yet been pursued by the telescope of the observer until sufficient data have been obtained to apply the process of analysis. This done, a few lines written upon paper indicate the precise times of their reappearance. These results, when first obtained, were so striking, and apparently so far beyond the reach of science itself, as almost to need the verification of experience. At the appointed times, however, the comets reappear, and science is thus verified by observation.

Comets: What mathematics proves in regard to them. Results striking. Verification.

§ 348. The great temple of nature is only to be opened by the keys of mathematical science. We may perhaps reach the vestibule, and gaze with wonder on its gorgeous exterior and its exact proportions, but we cannot open the portal and explore the apartments unless we use the appointed means. Those means are the exact sciences, which can only be acquired by discipline and severe mental labor.

Nature cannot be investigated without mathematics. Illustration.

The precious metals are not scattered profusely over the surface of the earth; they are, for wise purposes, buried in its bosom, and can be disinterred only by toil and labor. So with science: it comes not by inspiration; it is not borne to us on the wings of the wind; it can neither be extorted by power, nor purchased by wealth; but is the sure reward of diligent and assiduous labor. Is it worth that labor? What is it not worth? It has perforated the earth, and she has yielded up her treasures; it has guided in safety the bark of commerce over distant oceans, and brought to civilized man the treasures and choicest products of the remotest climes. It has scaled the heavens, and searched out the hidden laws which regulate and govern the material universe; it has travelled from planet to planet, measuring their magnitudes, surveying their surfaces, determining their days and nights, and the lengths of their seasons. It has also pushed its inquiries into regions of space, where it was supposed that the mind of the Omnipotent never yet had energized, and there located unknown worlds—calculating their diameters, and their times of revolution.

Science: Only to be acquired by study: It is worth study. What it has done for the wants of man: What it has done to make us acquainted with the universe.

§ 349. Mathematical science is a magnetic telegraph, which conducts the mind from orb

How mathematics

aid the mind in its inquiries: to orb through the entire regions of measured space. It enables us to weigh, in the balance of universal gravitation, the most distant planet of the heavens, to measure its diameter, to determine its times of revolution about a common centre and about its own axis, and to claim it as a part of our own system.

How they enlarge it: In these far reachings of the mind, the imagination has full scope for its highest exercise. It is not led astray by the false ideal and fed by illusive visions, which sometimes tempt reason from her throne, but is ever guided by the deductions of science; and its ideal and the real are united by the fixed laws of eternal truth. May be relied on.

Mind delights in certainty. § 350. There is that within us which delights in certainty. The mists of doubt obscure the mental, as the mists of the morning do the physical vision. We love to look at nature through a medium perfectly transparent, and to see every object in its exact proportions. The science of mathematics is that medium through which the mind may view, and thence understand all the parts of the physical universe. It makes manifest all its laws, discovers its wonderful harmonies, and displays the wisdom and omnipotence of the Creator. Why mathematics afford it.

## CHAPTER III.

### THE UTILITY OF MATHEMATICS CONSIDERED AS FURNISHING THOSE RULES OF ART WHICH MAKE KNOWLEDGE PRACTICALLY EFFECTIVE.

§ 351. THERE is perhaps no word in the English language less understood than PRACTICAL. By many it is regarded as opposed to theoretical. It has become a pert question of our day, "Whether such a branch of knowledge is practical?" "If any *practical* good arises from pursuing such a study?" "If it be not full time that old tomes be permitted to remain untouched in the alcoves of the library, and the minds of the young fed with the more stimulating food of modern progress?"

Practical: Little understood. Its popular signification. Questions relating to studies and books.

§ 352. Such inquiries are not to be answered by a taunt. They must be met as grave questions, and considered and discussed with calmness. They have possession of the public mind; they affect the foundations of education; they

Inquiries: How to be considered. Their influence.

Their importance. influence and direct the first steps; they control the very elements from which must spring the systems of public instruction.

Practical: Common acceptation: § 353. The term "practical," in its common acceptation, that is, in the sense in which it is often used, refers to the acquisition of useful knowledge by a short process. It implies a substitution of natural sagacity and "mother wit" for the results of hard study and laborious effort. It implies the use of knowledge before its acquisition; the substitution of the results of mere experiment for the deductions of science, and the placing of empiricism above philosophy.

What it implies.

In this sense, it is opposed to progress: § 354. In this view, the practical is adverse to sound learning, and directly opposed to real progress. If adopted, as a basis of national education, it would shackle the mind with the iron fetters of mere routine, and chain it down to the drudgery of unimproving labor. Under such a system, the people would become imitators and rule-men. Great and original principles would be lost sight of, and the spirit of investigation and inquiry would find no field for its legitimate exercise.

Consequences.

Right signification. But give to "practical" its true and right signification, and it becomes a word of the choicest import. In its right sense, it is the best means of making the true ideal the actual; that is, the best means of carrying into the business and practical affairs of life the conceptions and deductions of science. All that is truly great in the practical, is but the actual of an antecedent ideal.

Best means of applying knowledge.

§ 355. It is under this view that we now propose to consider the practical advantages of mathematical science. In the two preceding chapters we have pointed out its value as a means of mental development, and as affording facilities for the acquisition of knowledge. We shall now show how intimately it is blended with the every-day affairs of life, and point out some of the agencies which it exerts in giving practical development to the conceptions of the mind.

Mathematical science:

Its practical value.

§ 356. We begin with Arithmetic, as this branch of mathematics enters more or less into all the others. And what shall we say of its practical utility? It is at once an evidence and element of civilization. By its aid the child in the nursery numbers his toys, the housewife keeps her daily accounts, and the merchant sums up his daily business. The ten little characters,

Arithmetic considered practically.

which we call figures, thus perform a very important part in human affairs. They are sleepless sentinels watching over all the transactions of trade and commerce, and making known their final results. They superintend the entire business affairs of the world. Their daily records exhibit the results on the stock exchange, and of enterprises reaching over distant seas. The mechanic and artisan express the final results of all their calculations in figures. The dimensions of buildings, their length, breadth, and height, as well as the proportions of their several parts, are all expressed by figures before the foundation stones are laid; and indeed, all the results of science are reduced to figures before they can be made available in practice.

What figures do. Their value. Used by the mechanic: In building. Aid science.

§ 357. The rules and practice of all the mechanic arts are but applications of mathematical science. The mason computes the quantity of his materials by the principles of Geometry and the rules of Arithmetic. The carpenter frames his building, and adjusts all its parts, each to the others, by the rules of practical Geometry. The millwright computes the pressure of the water, and adjusts the driving to the driven wheel, by rules evolved from the formulas of analysis.

Mathematics useful in the mechanic arts. Examples.

§ 358. Workshops and factories afford marked illustrations of the utility and value of practical science. Here the most difficult problems are resolved, and the power of mind over matter exhibited in the most striking light. To the uninstructed eye of a casual observer, confusion appears to reign triumphant. But all the parts of that complicated machinery are adjusted to each other, and were indeed so arranged, and according to a general plan, before a single wheel was formed by the hand of the forger. The power necessary to do the entire work was first carefully calculated, and then distributed throughout the ramifications of the machinery. Each part was so arranged as to fulfil its office. Every circumference, and band, and cog, has its specific duty assigned it. The parts are made at different places, after patterns formed by the rules of science, and when brought together, fit exactly. They are but formed parts of an entire whole, over which, at the source of power, an ingenious contrivance, called the Governor, presides. His function is to regulate the force which shall drive the whole according to a uniform speed. He is so intelligent, and of such delicate sensibility, that on the slightest increase of velocity, he diminishes the force, and adds additional power the moment the speed

Workshops and factories exhibit applications of science.

Parts adjusted on a general plan.

Power calculated and distributed.

Parts fit in their proper places.

Governor:

Its functions.

All is but the result of science.

slackens. All this is the result of mathematical calculation. When the curious shall visit these exhibitions of ingenuity and skill, let them not suppose that they are the results of chance and experiment. They are the embodiments, by intelligent labor, of the most difficult investigations of mathematical science.

§ 359. Another striking example of the application of the principles of science is found in the steamship.

Steamship:

In the first place, the formation of her hull, so as to divide the waters with the least resistance, and at the same time receive from them the greatest pressure as they close behind her, is not an easy problem. Her masts are all to be set at the proper angle, and her sails so adjusted as to gain a maximum force. But the complication of her machinery, unless seen through the medium of science, baffles investigation, and exhibits a startling miracle. The burning furnace, the immense boilers, the massive cylinders, the huge levers, the pipes, the lifting and closing valves, and all the nicely-adjusted apparatus, appear too intricate to be comprehended by the mind at a single glance. Yet in all this complication—in all this variety of principle and workmanship, science has ex-

How the hull is formed.

Her masts: How adjusted.

Machinery:

The whole constructed

erted its power. There is not a cylinder, whose dimensions were not measured — not a lever, whose power was not calculated—nor a valve, which does not open and shut at the appointed moment. There is not, in all this structure, a bolt, or screw, or rod, which was not provided for before the great shaft was forged, and which does not bear to that shaft its proper proportion. And when the workmanship is put to the test, and the power of steam is urging the vessel on her distant voyage, science alone can direct her way.

according to the principles of science: From a general plan. By what means navigated:

In the captain's cabin are carefully laid away, for daily use, maps and charts of the port which he leaves, of the ocean he traverses, and of the coasts and harbors to which he directs his way. On these are marked the results of much scientific labor. The shoals, the channels, the points of danger and the places of security, are all indicated. Near by, hangs the barometer, constructed from the most abstruse mathematical formulas, to indicate changes in the weight of the atmosphere, and admonish him of the approaching tempest. On his table lie the sextant, and the tables of Bowditch. These enable him, by observations on the heavenly bodies, to mark his exact place on the chart, and learn his position on the surface of the earth. Thus, practical

Her charts: Their contents and uses. Barometer: Sextant: Their uses.

Science guides the ship: science, which shaped the keel of the ship to its proper form, and guided the hand of the mechanic in every workshop, is, under Providence, the means of conducting her in safety over the ocean. It is, indeed, the cloud by day and the pillar of fire by night. Guiding the bark of commerce over trackless waters, it brings distant lands into proximity, and into political and social relations.

What thus accomplishes.

Illustration.

"We have before us an anecdote communicated to us by a naval officer,* distinguished for the extent and variety of his attainments, which shows how impressive such results may become in practice. He sailed from San Blas, on the west coast of Mexico, and after a voyage of eight thousand miles, occupying eighty-nine days, arrived off Rio de Janeiro; having in this interval passed through the Pacific Ocean, rounded Cape Horn, and crossed the South Atlantic, without making any land, or even seeing a single sail, with the exception of an American whaler off Cape Horn. Arrived within a week's sail of Rio, he set seriously about determining, by lunar observations, the precise line of the ship's course, and its situation in it, at a determinate moment; and having ascertained this within

Capt. Hall's voyage.

Its length:

and incidents.

Observations taken.

* Captain Basil Hall.

from five to ten miles, ran the rest of the way by those more ready and compendious methods, known to navigators, which can be safely employed for short trips between one known point and another, but which cannot be trusted in long voyages, where the moon is the only sure guide.

Remarkable coincidence.

Short methods.

"The rest of the tale, we are enabled, by his kindness, to state in his own words: 'We steered towards Rio de Janeiro for some days after taking the lunars above described, and having arrived within fifteen or twenty miles of the coast, I hove-to at four in the morning, till the day should break, and then bore up: for although it was very hazy, we could see before us a couple of miles or so. About eight o'clock it became so foggy, that I did not like to stand in further, and was just bringing the ship to the wind again, before sending the people to breakfast, when it suddenly cleared off, and I had the satisfaction of seeing the great Sugar-Loaf Rock, which stands on one side of the harbor's mouth, so nearly right ahead that we had not to alter our course above a point in order to hit the entrance of Rio. This was the first land we had seen for three months, after crossing so many seas, and being set backwards and forwards by innumerable currents and foul winds.' The effect on all on board might well be conceived to have been electric;

Particulars stated.

Arrival at Rio.

Discovery of Harbor.

First land in three months.

Effect

on the crew. and it is needless to remark how essentially the authority of a commanding officer over his crew may be strengthened by the occurrence of such incidents, indicative of a degree of knowledge and consequent power beyond their reach."*

Surveying. § 360. A useful application of mathematical science is found in the laying out and measurement of land. Measurement of land. The necessity of such measurement, and of dividing the surface of the earth into portions, gave rise to the science of Geometry. Ownership: How determined. The ownership of land could not be determined without some means of running boundary lines, and ascertaining limits. Levelling is also connected with this branch of practical mathematics.

By the aid of these two branches of practical science, we measure and determine the area or Contents of ground. contents of ground; make maps of its surface; measure the heights of hills and mountains; Rivers. find the directions of rivers; measure their volumes, and ascertain the rapidity of their currents. So certain and exact are the results, that entire countries are divided into tracts of convenient size, and the rights of ownership fully Certainty. secured. The rules for mapping, and the con-

* Sir John Herschel, on the study of Natural Philosophy.

ventional methods of representing the surface of ground, the courses of rivers, and the heights of mountains, are so well defined, that the natural features of a country may be all indicated on paper. Thus, the topographical features of all the known parts of the earth may be correctly and vividly impressed on the mind, by a map, drawn according to the rules of art, by the human hand.

Mapping.

Features of the ground.

Their representation.

§ **361.** Our own age has been marked by a striking application of science, in the construction of railways. Let us contemplate for a moment the elements of the problem which is presented in the enterprise of constructing a railroad between two given points.

Railways.

The problem presented.

In the first place, the route must be carefully examined to ascertain its general practicability. The surveyor, with his instruments, then ascertains all the levels and grades. The engineer examines these results to determine whether the power of steam, in connection with the best combination of machinery, will enable him to overcome the elevations and descend the declivities in safety. He then calculates the curves of the road, the excavations and fillings, the cost of the bridges and the tunnels, if there are any; and then adjusts the steam-power to meet

Examination of their routes.

Surveys.

Office of the engineer.

Calculations of curves.

Completion and use.

the conditions. In a few months after the enterprise is undertaken, the locomotive, with its long train of passenger and freight cars, rushes over the tract with a superhuman power, and fulfils the office of uniting distant places in commercial and social relations.

The striking fact.

But that which is most striking in all this, is the fact, that before a stump is grubbed, or a spade put into the ground, the entire plan of the work, having been subjected to careful analysis, is fully developed in all its parts. The construction is but the actual of that perfect ideal which the mind forms within itself, and which can spring only from the far-reaching and immutable principles of abstract science.

The whole the result of science.

§ 362. Among the most useful applications of practical science, in the present century, is the introduction of the Croton water into the city of New York.

Croton aqueduct.

In the Highlands of the Hudson, about fifty miles from the city, the gushing springs of the mountains indicate the sources of the Croton river, which enters the Hudson a few miles below Peekskill. At a short distance from the mouth, a dam fifty-five feet in height is thrown across the river, creating an artificial lake for the permanent supply of water. The area of this

Sources of the river.

Principal reservoir.

lake is equal to about four hundred acres. The aqueduct commences at the Croton dam, on a line forty feet above the level of the Hudson river, and runs, as near as the nature of the ground will permit, along the east bank, till it reaches its final destination in the reservoirs of the city. There are on the line sixteen tunnels, varying in length from 160 to 1,263 feet, making an aggregate length of 6,841 feet. The heights of the ridges above the grade level of the tunnels range from 25 to 75 feet. Twenty-five streams are crossed by the aqueduct in Westchester county, varying from 12 to 70 feet below the grade line, and from 25 to 83 feet below the top covering of the aqueduct. The Harlem river is passed at an elevation of 120 feet above the surface of the water. The average dimensions of the interior of the aqueduct, are about seven feet in width and eight feet in height.

Its area. Aqueduct. Its tunnels: Their heights. Streams crossed. Harlem river:

The width of the Harlem river, at the point where the aqueduct crosses it, is six hundred and twenty feet, and the general plan of the bridge is as follows: There are eight arches, each of 80 feet span, and seven smaller arches, each of 50 feet span, the whole resting on piers and abutments. The length of the bridge is 1,450 feet. The height of the river piers from the lowest foundation is 96 feet. The arches

Its width. Bridge: Its length:

Its height: are semi-circular, and the height from the lowest foundation of the piers to the top of the Its width. parapet is 149 feet. The width across, on the top, is 21 feet.

To afford a constant supply of water for distribution in the city two large reservoirs have Receiving Reservoir: been constructed, called the receiving reservoir and the distributing reservoir. The surface of the receiving reservoir, at the water-line, is equal Its extent. to thirty-one acres. It is divided into two parts by a wall running east and west. The depth of Depth of water. water in the northern part is twenty feet, and in the southern part thirty feet.

Distributing Reservoir: The distributing reservoir is located on the highest ground which adjoins the city, known Its capacity. as Murray Hill. The capacity of this reservoir is equal to 20,000,000 of gallons, which is about one-seventh that of the receiving reservoir, and the depth of water is thirty-six feet.

Power of science. The full power of science has not yet been illustrated. A perfect plan of this majestic structure was arranged, or should have been, before a stone was shaped, or a pickaxe put into the ground. The complete conception, by a single mind, of its general plan and minutest details, was necessary to its successful prosecu- What it accomplished. tion. It was within the range and power of science to have given the form and dimensions of every stone, so that each could have been shaped at the quarry. The parts are so connected by the laws of the geometrical forms, that the dimensions and shape of each stone was exactly determined by the nature of that portion of the structure to which it belonged.

Connections of the parts.

§ 363. We have presented this outline of the Croton aqueduct mainly for the purpose of illustrating the power and celebrating the triumphs of mathematical science. High intellect, it is true, can alone use the means in a work so complicated, and embracing so great a variety of intricate details. But genius, even of the highest order, could not accomplish, without continued trial and laborious experiment, such an undertaking, unless strengthened and guided by the immutable truths of mathematical science.

View of the Croton aqueduct: Why given.

Little accomplished without science.

§ 364. The examination of this work cannot but fill the mind with a proud consciousness of the power and skill of man. The struggling brooks of the mountains are collected together—accumulated—conducted for forty miles through a subterranean channel, to form small lakes in the vicinity of a populous city.

What science has done.

From these sources, by an unseen process, the

pure water is carried to every dwelling in the large metropolis. The turning of a faucet de-

Consequences which have followed.

livers it from a spring at the distance of fifty miles, as pure as when it gushes from its granite hills. That unseen power of pressure, which resides in the fluid as an organic law, exerts its force with unceasing and untiring energy. To minds enlightened by science, and skill directed by its rules, we are indebted for one of the noblest works of the present century. May we

Conclusion.

not, therefore, conclude that science is the only sure means of giving practical development to those great conceptions which confer lasting benefits on mankind? "All that is truly great in the practical, is but the result of an antecedent ideal."

# APPENDIX.

A COURSE OF MATHEMATICS—WHAT IT SHOULD BE.

§ 365. A COURSE of mathematics should present the outlines of the science, so arranged, explained, and illustrated as to indicate all those general methods of application, which render it effective and useful. This can best be done by a series of works embracing all the topics, and in which each topic is separately treated.

A course of Mathematics.

§ 366. Such a series should be formed in accordance with a fixed plan; should adopt and use the same terms in all the branches; should be written throughout in the same style; and present that entire unity which belongs to the subject itself.

How it should be formed.

Unity of the subject.

§ 367. The reasonings of mathematics and the processes of investigation, are the same in

Reasonings

the same in all branches. every branch, and have to be learned but once, if the same system be studied throughout. The Different kinds of notation. different kinds of notation, though somewhat unlike in the different subjects of the science, are, in fact, but dialects of a common language.

Language need be learned but once. § 368. If, then, the language is, or may be made essentially the same in all the branches of mathematical science; and if there is, as has been fully shown, no difference in the processes In what consists the difficulty? of reasoning, wherein lies that difficulty in the acquisition of mathematical knowledge which is often experienced by students, and whence the origin of that opinion that the subject itself is dry and difficult?

A general law, if known, renders a subject easy. § 369. Just in proportion as a branch of knowledge is compactly united by a common law, is the facility of acquiring that knowledge, if we observe the law, and the difficulty of acquiring Faculties required in mathematics. it, if we pay no attention to the law. The study of mathematics demands, at every step, close attention, nice discrimination, and certain judgment. These faculties can only be developed How first cultivated: by culture. They must, like other faculties, pass through the states of infancy, growth, and maturity. They must be first exercised on sensible and simple objects; then on elementary abstract ideas; and finally, on generalizations and the higher combinations of thought in the pure ideal. On what finally exercised.

§ 370. Have educators fully realized that the first lessons in numbers impress the first elements of mathematical science? that the first connections of thought which are there formed become the first threads of that intellectual warp which gives tone and strength to the mind? Have they yet realized that every process is, or should be, like the stone of an arch, formed to fill, in the entire structure, the exact place for which it is designed? and that the unity, beauty, and strength of the whole depend on the adaptation of the parts to each other? Have they sufficiently reflected on the confusion which must arise from attempting to put together and harmonize different parts of discordant systems? to blend portions that are fragmentary, and to unite into a placid and tranquil stream trains of thought which have not a common source?

Arithmetic the most important branch.

All the subjects connected.

Necessity of unity in all the parts.

§ 371. Some have supposed that Arithmetic may be well taught and learned without the aid of a text-book; or, if studied from a book, that the teacher may advantageously substitute his own methods for those of the author, inasmuch

A text-book

to be followed.

as such substitution is calculated to widen the field of investigation, and excite the mind of the pupil to new inquiries.

Reasons.

Admitting that every teacher of reasonable intelligence, will discover methods of communicating instruction better adapted to the peculiarities of his own mind, than all the methods employed by the author he may use; will it be safe, as a general rule, to *substitute* extemporaneous methods for those which have been subjected to the analysis of science and the tests of experience? Is it safe to substitute the results of known laws for conjectural judgments? But if they are as good, or better even, as isolated processes, will they answer as well, in their new places and connections, as the parts rejected? Will the balance-wheel of a chronometer give as steady a motion to a common watch as the more simple and less perfect contrivance to which all the other parts are adapted?

Even a better method, when substituted, may not harmonize with the other parts of the work.

Illustration.

One of the reasons why mathematics is difficult.

§ 372. If these questions have significance, we have found at least one of the causes that have impeded the advancement of mathematical science, viz. the attempt to unite in the same course of instruction fragments of different systems; thus presenting to the mind of the learner the same terms differently defined, and the same principles differently explained, illustrated, and applied. It is mutual relation and *connection* which bring sets of facts under general laws; it is mutual relation and connection of ideas which form a process of science; it is the mutual connection and relation of such processes which constitute science itself.

Connection very important.

§ 373. I would by no means be understood as expressing the opinion that a student or teacher of mathematics should limit his researches to a single author; for, he must necessarily read and study many. I speak of the pupil alone, who *must be taught one method at a time, and taught that well, before he is able to compare different methods with each other.*

A teacher should read many books, and teach one system.

ORDER OF THE SUBJECTS—ARITHMETIC.

§ 374. Arithmetic is the most useful and simple branch of mathematical science, and is the first to be taught. If, however, the pupil has time for a full course, I would by no means recommend him to finish his Arithmetic before studying a portion of Algebra.

Arithmetic:

Connection with Algebra.

## ALGEBRA.

Algebra: § 375. Algebra is but a universal Arithmetic, with a more comprehensive notation. Its elements are acquired more readily than the higher and hidden properties of numbers; and indeed, the elements of any branch of mathematics are more simple than the higher principles of the preceding subject; so that all the subjects can best be studied in connection with those which precede and follow.

How it should be studied:

Should precede Geometry: Why. When Geometry should be commenced.

§ 376. Algebra, in a regular course of instruction, should precede Geometry, because the elementary processes do not require, in so high a degree, the exercise of the faculties of abstraction and generalization. But when we have completed the equation of the second degree, the processes become more difficult, the abstractions more perfect, and the generalizations more extended. Here then I would pause and commence Geometry.

## GEOMETRY.

Geometry. § 377. Geometry, as one of the subjects of mathematical science, has been fully considered in Book II. It is referred to here merely to mark its place in a regular course of instruction.

## TRIGONOMETRY—PLANE AND SPHERICAL.

§ 378. The next subject in order, after Geometry, is Trigonometry: a mere application of the principles of Arithmetic, Algebra, and Geometry to the determination of the sides and angles of triangles. As triangles are of two kinds, viz. those formed by straight lines and those formed by the arcs of great circles on the surface of a sphere; so Trigonometry is divided into two parts: Plane and Spherical. Plane Trigonometry explains the methods, and lays down the necessary rules for finding the remaining sides and angles of a plane triangle, when a sufficient number are known or given. Spherical Trigonometry explains like processes, and lays down similar rules for spherical triangles.

Trigonometry: What it is. Two kinds. Plane. Spherical.

## SURVEYING AND LEVELLING.

§ 379. The application of the principles of Trigonometry to the measurement of portions of the earth's surface, is called Surveying; and similar applications of the same principles to the determination of the difference between the distances of any two points from the centre of the earth, is called Levelling. These subjects, which follow Trigonometry, not only embrace the va-

Surveying. Levelling.

What they embrace. rious methods of calculation, but also a description of the necessary Instruments and Tables. They should be studied immediately after Trigonometry; of which, indeed, they are but applications.

## DESCRIPTIVE GEOMETRY.

Descriptive Geometry: § 380. Descriptive Geometry is that branch of mathematics which considers the positions of the geometrical magnitudes, as they may exist in space, and determines these positions by referring the magnitudes to two planes called the Planes of Projection.

Its nature. It is, indeed, but a development of those general methods, by which lines, surfaces, and solids may be presented to the mind by means of drawings made upon paper. The processes of this development require the constant exercise of the conceptive faculty. All geometrical magnitudes may be referred to two planes of projection, and their representations on these planes will express to the mind, their forms, extent, and also their positions or places in space. From these representations, the mind perceives, as it were, at a single view, the magnitudes themselves, as they exist in space; traces their boundaries, measures their extent, and sees all their parts separately and in their connection.

What its study accomplishes.

How.

In France, Descriptive Geometry is an important element of education. It is taught in most of the public schools, and is regarded as indispensable to the architect and engineer. It is, indeed, the only means of so reducing to paper, and presenting at a single view, all the complicated parts of a structure, that the drawing or representation of it can be read at a glance, and all the parts be at once referred to their appropriate places. It is to the engineer or architect not only a general language by which he can record and express to others all his conceptions, but is also the most powerful means of extending those conceptions, and subjecting them to the laws of exact science.

How regarded in France.

Its value as a practical branch.

### SHADES, SHADOWS, AND PERSPECTIVE.

§ 381. The application of Descriptive Geometry to the determination of shades and shadows, as they are found to exist on the surfaces of bodies, is one of the most striking and useful applications of science; and when it is further extended to the subject of Perspective, we have all that is necessary to the exact representation of objects as they appear in nature. An accurate perspective and the right distribution of light and shade are the basis of every work of

Shades, Shadows, and Perspective.

Their use. the fine arts. Without them, the sculptor and the painter would labor in vain: the chisel of Canova would give no life to the marble, nor the touches of Raphael to the canvas.

## ANALYTICAL GEOMETRY.

Analytical Geometry.

§ 382. Analytical Geometry is the next subject in a regular course of mathematical study, though it may be studied before Descriptive Geometry. The importance of this subject cannot be exaggerated. In Algebra, the symbols of quantity have generally so close a connection with numbers, that the mind scarcely realizes the extent of the generalization; and the power of analysis, arising from the changes that may take place among the quantities which the symbols represent, cannot be fully explained and developed.

Its importance: Valuable as a study.

But in Analytical Geometry, where all the magnitudes are brought under the power of analysis, and all their properties developed by the combined processes of Algebra and Geometry, we are brought to feel the extent and potency of those methods which combine in a single equation every discovered and undiscovered property of every line, straight or curved, which can be formed by the intersection of a cone and plane.

Reasons.

To develop every property of the Conic Sections from a single equation, and that an equation only of the second degree, by the known processes of Algebra, and thus interpret the results, is a far different exercise of the mind from that which arises from searching them out by the tedious and disconnected methods of separate propositions. The first traces all from an inexhaustible fountain by the known laws of analytical investigation, applicable to all similar cases, while the latter adopts particular processes applicable to special cases only, without any general law of connection.

Its extent.

Its methods of operation.

## DIFFERENTIAL AND INTEGRAL CALCULUS.

§ 383. The Differential and Integral Calculus presents a new view of the power, extent, and applications of mathematical science. It should be carefully studied by all who seek to make high attainments in mathematical knowledge, or who desire to read the best works on Natural and Experimental Philosophy. It is that field of mathematical investigation, where genius may exert its highest powers and find its most certain rewards.

Differential and Integral Calculus.

What persons should study it.

# INDEX.

THE END.

# CHAMBERS' EDUCATIONAL COURSE.

## THE SCIENTIFIC SECTION.

The Messrs. Chambers have employed the first professors in Scotland in the preparation of these works. They are now offered to the schools of the United States, under the American revision of D. M. REESE, M. D., LL. D., *late Superintendent of Public Schools in the city and county of New York.*

I. **CHAMBERS' TREASURY OF KNOWLEDGE.**
II. **CLARK'S ELEMENTS OF DRAWING AND PERSPECTIVE.**
III. **CHAMBERS' ELEMENTS OF NATURAL PHILOSOPHY.**
IV. **REID & BAIN'S CHEMISTRY AND ELECTRICITY.**
V. **HAMILTON'S VEGETABLE AND ANIMAL PHYSIOLOGY**
VI. **CHAMBERS' ELEMENTS OF ZOOLOGY.**
VII. **PAGE'S ELEMENTS OF GEOLOGY.**

---

"It is well known that the original publishers of these works (the Messrs. Chambers of Edinburgh) are able to command the best talent in the preparation of their books, and that it is their practice to deal faithfully with the public. This series will not disappoint the reasonable expectations thus excited. They are elementary works prepared by authors in every way capable of doing justice to their respective undertakings, and who have evidently bestowed upon them the necessary time and labor to adapt them to their purpose. We recommend them to teachers and parents with confidence. If not introduced as class-books in the school, they may be used to excellent advantage in general exercises, and occasional class exercises, for which every teacher ought to provide himself with an ample store of materials. The volumes may be had separately; and the one first named, in the hands of a teacher of the younger classes, might furnish an inexhaustible fund of amusement and instruction. Together, they would constitute a rich treasure to a family of intelligent children, and impart a thirst for knowledge."—*Vermont Chronicle.*

---

"Of all the numerous works of this class that have been published, there are none that have acquired a more thoroughly deserved and high reputation than this series. The Chambers, of Edinburgh, well known as the careful and intelligent publishers of a vast number of works of much importance in the educational world, are the fathers of this series of books, and the American editor has exercised an unusual degree of judgment in their preparation for the use of schools as well as private families in this country."—*Philad. Bulletin.*

---

"The titles furnish a key to the contents, and it is only necessary for us to say, that the material of each volume is admirably worked up, presenting with sufficient fulness and with much clearness of method the several subjects which are treated."—*Cin Gazette.*

---

"We notice these works, not merely because they are *school books*, but for the purpose of expressing our thanks, as the 'advocate' of the educational interests of the people and their children, to the enterprising publishers of these and many other valuable works of the same character, the tendency of which is to diffuse useful knowledge throughout the masses, for the good work they are doing, and the hope that their reward may be commensurate with their deserts."—*Maine School Advocate.*

*Davies' System of Mathematics.*

# MATHEMATICAL WORKS,

IN A SERIES OF THREE PARTS:

## ARITHMETICAL, ACADEMICAL, AND COLLEGIATE.

BY CHARLES DAVIES, L.L.D

### I. THE ARITHMETICAL COURSE FOR SCHOOLS.

1. PRIMARY TABLE-BOOK.
2. FIRST LESSONS IN ARITHMETIC.
3. SCHOOL ARITHMETIC. (Key separate.)
4. GRAMMAR OF ARITHMETIC.

### II. THE ACADEMIC COURSE.

1. THE UNIVERSITY ARITHMETIC. (Key separate.)
2. PRACTICAL GEOMETRY AND MENSURATION.
3. ELEMENTARY ALGEBRA. (Key separate.)
4. ELEMENTARY GEOMETRY.
5. ELEMENTS OF SURVEYING.

### III. THE COLLEGIATE COURSE.

1. DAVIES' BOURDON'S ALGEBRA.
2. DAVIES' LEGENDRE'S GEOMETRY AND TRIGONOMETRY.
3. DAVIES' ANALYTICAL GEOMETRY.
4. DAVIES' DESCRIPTIVE GEOMETRY.
5. DAVIES' SHADES, SHADOWS, AND PERSPECTIVE.
6. DAVIES' DIFFERENTIAL AND INTEGRAL CALCULUS.

**DAVIES' LOGIC AND UTILITY OF MATHEMATICS.**

This series, combining all that is most valuable in the various methods of European instruction, improved and matured by the suggestions of more than thirty years' experience, now forms the only complete consecutive course of Mathematics. Its methods, harmonizing as the works of one mind, carry the student onward by the same analogies and the same laws of association, and are calculated to impart a comprehensive knowledge of the science, combining clearness in the several branches, and unity and proportion in the whole. Being the system so long in use at West Point, through which so many men, eminent for their scientific attainments, have passed, and having been adopted, as Text Books, by most of the colleges in the United States, it may be justly regarded as our

NATIONAL SYSTEM OF MATHEMATICS.

*Parker's Natural Philosophy.*

# NATURAL AND EXPERIMENTAL PHILOSOPHY,

## FOR SCHOOLS AND ACADEMIES.

BY R. G. PARKER, A. M.,

*Author of "Rhetorical Reader," "Exercises in English Composition," "Outlines of History," etc., etc.*

**I. PARKER'S JUVENILE PHILOSOPHY.**

**II. PARKER'S FIRST LESSONS IN NATURAL PHILOSOPHY.**

**III. PARKER'S SCHOOL COMPENDIUM OF PHILOSOPHY.**

The use of school apparatus for illustrating and exemplifying the principles of Natural and Experimental Philosophy, has, within the last few years, become so general as to render necessary a work which should combine, in the same course of instruction, the theory, with a full description of the apparatus necessary for illustration and experiment. The work of Professor Parker, it is confidently believed, fully meets that requirement. It is also very full in the general facts which it presents—clear and concise in its style—and entirely scientific and natural in its arrangement.

"This work is better adapted to the present state of natural science than any other similar production with which we are acquainted."—*Wayne Co. Whig.*

"This is a school-book of no mean pretensions and no ordinary value."—*Albany Spectator.*

"We predict for this valuable and beautifully-printed work the utmost success."—*Newark Daily Advertiser.*

"The present volume strikes us as having very marked merit."—*N. Y. Courier.*

"It seems to me to have hit a happy medium between the too simple and the two abstract."—*B. A. Smith, Principal of Leicester Academy, Mass.*

"I have no hesitation in saying that Parker's Natural Philosophy is the most valuable elementary work I have seen."—*Gilbert Langdon Hume, Prof. Nat. Phil. N. Y. City.*

"I am happy to say that Parker's Philosophy will be introduced and adopted in 'Victoria College,' at the commencement of the next collegiate year in autumn; and I hope that will be but the commencement of the use of so valuable an elementary work in our schools in this country. The small work of Parker's (Parker's First Lessons) was introduced the last term in a primary class of the institution referred to, and that with great success. I intend to recommend its use shortly into the *model* school in this city, and the larger work to the students of the provincial Normal School."—*E. Ryerson, Superintendent of Public Instruction of Upper Canada.*

"I have examined Parker's First Lessons and Compendium of Natural and Experimental Philosophy, and am much pleased with them. I have long felt dissatisfaction with the Text-Books on this subject most in use in this section, and am happy now to find books that I can recommend. I shall introduce them immediately into my school." *Hiram Orcutt, Principal of Thetford Academy, Vermont.*

"I have no hesitation in pronouncing it the *best work on the subject now published.* We shall use it here, and I have already secured its adoption in some of the high-schools and academies in our vicinity."—*M. D. Leggett, Sup. of Warren Public Schools.*

"We are glad to see this little work on natural philosophy, because the amount of valuable information under all these heads, to be gained from it by any little boy or girl, is inestimable. It puts them, too, upon the right track after knowledge, and prevents their minds from being weakened and wasted by the sickly sentimentality of tales, novels, and poetry, which will always occupy the attention of the mind when nothing more useful has taken possession of it."—*Mississippian.*

# MRS. EMMA WILLARD'S

## SERIES OF SCHOOL HISTORIES AND CHARTS.

I. **WILLARD'S HISTORY OF THE UNITED STATES, OR REPUBLIC OF AMERICA.** 8vo. Price $1.50.

II. **WILLARD'S SCHOOL HISTORY OF THE UNITED STATES.** 63 cts.

III. **WILLARD'S AMERICAN CHRONOGRAPHER.** $1.50.

---

I. **WILLARD'S UNIVERSAL HISTORY IN PERSPECTIVE.** $1.50.

II. **WILLARD'S TEMPLE OF TIME.** Mounted, $1.25. Bound, 75 cts

III. **WILLARD'S HISTORIC GUIDE.** 50 cts.

IV. **WILLARD'S ENGLISH CHRONOGRAPHER.**

### WILLARD'S UNITED STATES.

The *Hon. Dan. Webster* says, of an early edition of the above work, in a letter to the author, "I KEEP IT NEAR ME, AS A BOOK OF REFERENCE, ACCURATE IN FACTS AND DATES."

---

"THE COMMITTEE ON BOOKS OF THE WARD SCHOOL ASSOCIATION RESPECTFULLY REPORT:

"That they have examined Mrs. Willard's History of the United States with peculiar interest, and are free to say, that it is in their opinion decidedly the best treatise on this interesting subject that they have seen. As a school-book, its proper place is among the first. The language is remarkable for simplicity, perspicuity, and neatness; youth could not be trained to a better taste for language than this is calculated to impart. It places at once, in the hands of American youth, the history of their country from the day of its discovery to the present time, and exhibits a clear arrangement of all the great and good deeds of their ancestors, of which they now enjoy the benefits, and inherit the renown. The struggles, sufferings, firmness, and piety of the first settlers are delineated with a masterly hand."—*Extract from a Report of the Ward School Teachers' Association of the City of New York.*

---

"We consider the work a remarkable one, in that it forms the best book for general reading and reference published, and at the same time has no equal, in our opinion, as a text-book. On this latter point, the profession which its author has so long followed with such signal success, rendered her peculiarly a fitting person to prepare a text-book."—*Boston Traveller.*

---

"MRS. WILLARD'S SCHOOL HISTORY OF THE UNITED STATES.—It is one of those rare things, a good school-book; infinitely better than any of the United States Histories fitted for schools, which we have at present."—*Cincinnati Gazette.*

---

"We think we are warranted in saying, that it is better adapted to meet the wants of our schools and academies in which history is pursued, than any other work of the kind now before the public. The style is perspicuous and flowing, and the prominent points of our history are presented in such a manner as to make a deep and lasting impression on the mind. We could conscientiously say much more in praise of this book, but must content ourselves by heartily commending it to the attention of those who are anxious to find a good text-book of American history for the use of schools."—*Newburyport Watchman.*

## PARKER'S RHETORICAL READER. 12mo.

Exercises in Rhetorical Reading, designed to familiarize readers with the pauses and other marks in general use, and lead them to the practice of modulation and inflection of the voice. By R. G. PARKER, author of "Exercises in English Composition," "Compendium of Natural Philosophy," &c., &c.

---

This work possesses many advantages which commend it to favor, among which are the following:—It is adapted to all classes and schools, from the highest to the lowest. It contains a practical illustration of all the marks employed in written language; also lessons for the cultivation, improvement, and strengthening of the voice, and instructions as well as exercises in a great variety of the principles of Rhetorical Reading, which cannot fail to render it a valuable auxiliary in the hands of any teacher. Many of the exercises are of sufficient length to afford an opportunity for each member of any class, however numerous, to participate in the same exercise—a feature which renders it convenient to examining committees. The selections for exercises in reading are from the most approved sources, possessing a salutary moral and religious tone, without the slightest tincture of sectarianism.

---

"I have to acknowledge the reception through your kindness of several volumes. I have not as yet found time to examine minutely all the books. Of Mr. Parker's Rhetorical Reader, however, I am prepared to speak in the highest terms. I think it so well adapted to the wants of pupils, that I shall introduce it immediately in the Academy of which I am about to take charge at Madison, in this state. It is the best thing of the kind I have yet found. I cannot say too much in its favor."—*John G. Clark, Rector of the Madison Male Academy, Athens, Ga.*

---

"Mr. Parker has made the public his debtor by some of his former publications—especially the 'Aids to English Composition'—and by this he has greatly increased the obligation. There are reading books almost without number, but very few of them pretend to give instructions how to read, and, unluckily, few of our teachers are competent to supply the defect. If young persons are to be taught to read well, it must generally be done in the primary schools, as the collegiate term affords too little time to begin and accomplish that work. We have seen no other 'Reader' with which we have been so well pleased; and as an evidence of our appreciation of its worth, we shall lay it aside for the use of a certain juvenile specimen of humanity in whose affairs we are specially interested."—*Christian Advocate.*

---

"We cannot too often urge upon teachers the importance of reading, as a part of education, and we regard it as among the auspicious signs of the times, that so much more attention is given, by the best of teachers, to the cultivation of a power which is at once a most delightful accomplishment, and of the first importance as a means of discipline and progress. In this work, Mr. Parker's volume, we are sure, will be found a valuable aid."—*Vermont Chronicle.*

---

"The title of this work explains its character and design, which are well carried out by the manner in which it is executed. As a class-book for students in elocution, or as an ordinary reading book, we do not think we have seen any thing superior. The distinguishing characteristic of its plan is to assume some simple and familiar example, which will be readily understood by the pupil, and which Nature will tell him how to deliver properly, and refer more difficult passages to this, as a model. There is, however, another excellence in the work, which we take pleasure in commending; it is the progressiveness with which the introductory lessons are arranged. In teaching every art and science this is indispensable, and in none more so than in that of elocution. The pieces for exercise in reading are selected with much taste and judgment. We have no doubt that those who use this book will be satisfied with its success."—*Teacher's Advocate.*

*Brooks's Greek and Latin Classics.*

# PROFESSOR BROOKS'S
# GREEK AND LATIN CLASSICS.

This series of the Greek and Latin Classics, by N. C. Brooks, of Baltimore, is on an improved plan, with peculiar adaptation to the wants of the American student. To secure accuracy of text in the works that are to appear, the latest and most approved European editions of the different classical authors will be consulted. Original illustrative and explanatory notes, prepared by the Editor, will accompany the text. These notes, though copious, will be intended to direct and assist the student in his labors, rather than by rendering every thing too simple, to supersede the necessity of due exertion on his own part, and thus induce indolent habits of study and reflection, and feebleness of intellect.

In the notes that accompany the text, care will be taken, on all proper occasions, to develop and promote in the mind of the student, sound principles of Criticism, Rhetoric, History, Political Science, Morals, and general Religion; so that he may contemplate the subject of the author he is reading, not within the circumscribed limits of a mere rendering of the text, but consider it in all its extended connections—and thus learn to *think*, as well as to *translate*.

## BROOKS'S FIRST LATIN LESSONS.

This is adapted to any Grammar of the language. It consists of a Grammar, Reader, and Dictionary combined, and will enable any one to acquire a knowledge of the elements of the Latin Language, without an instructor. It has already passed through five editions. 18mo.

## BROOKS'S CÆSAR'S COMMENTARIES. (*In press.*)

This edition of the Commentaries of Cæsar on the Gallic War, besides critical and explanatory notes embodying much information, of an historical, topographical, and military character, is illustrated by maps, portraits, views, plans of battles, &c. It has a good Clavis, containing all the words. Nearly ready. 12mo.

## BROOKS'S OVID'S METAMORPHOSES. 8vo.

This edition of Ovid is expurgated, and freed from objectionable matter. It is elucidated by an analysis and explanation of the fables, together with original English notes, historical, mythological, and critical, and illustrated by pictorial embellishments; with a Clavis giving the meaning of all the words with critical exactness. Each fable contains a plate from an original design, and an illuminated initial letter.

## BROOKS'S ECLOGUES AND GEORGICS OF VIRGIL. (*In press.*)

This edition of Virgil is elucidated by copious original notes, and extracts from ancient and modern pastoral poetry. It is illustrated by plates from original designs, and contains a Clavis giving the meaning of all the words. 8vo.

## BROOKS'S FIRST GREEK LESSONS. 12mo.

This Greek elementary is on the same plan as the Latin Lessons, and affords equal facilities to the student. The paradigm of the Greek verb has been greatly simplified and valuable exercises in comparative philology introduced.

## BROOKS'S GREEK COLLECTANEA EVANGELICA. 12mo.

This consists of portions of the Four Gospels in Greek, arranged in Chronological order; and forms a connected history of the principal events in the Saviour's life and ministry. It contains a Lexicon, and is illustrated and explained by notes.

## BROOKS'S GREEK PASTORAL POETS. (*In press.*)

This contains the Greek Idyls of Theocritus, Bion, and Moschus, elucidated by notes and copious extracts from ancient and modern pastoral poetry. Each Idyl is illustrated by beautiful plates from original designs. It contains a good Lexicon.

*Page's Theory and Practice of Teaching.*

**THEORY AND PRACTICE OF TEACHING;**

OR THE

# MOTIVES OF GOOD SCHOOL-KEEPING.

BY DAVID PAGE, A.M.,

LATE PRINCIPAL OF THE STATE NORMAL SCHOOL, NEW YORK.

"I received a few days since your 'Theory and Practice, &c.,' and a capital *theory* and capital *practice* it is. I have read it with unmingled delight. Even if I should look through a critic's microscope, I should hardly find a single sentiment to dissent from, and certainly not one to condemn. The chapters on *Prizes* and on *Corporal Punishment* are truly admirable. They will exert a most salutary influence. So of the views *sparsim* on moral and religious instruction, which you so earnestly and feelingly insist upon, and yet within true Protestant limits. IT IS A GRAND BOOK, AND I THANK HEAVEN THAT YOU HAVE WRITTEN IT."—*Hon. Horace Mann, Secretary of the Board of Education in Massachusetts.*

"Were it our business to examine teachers, we would never dismiss a candidate without naming this book. Other things being equal, we would greatly prefer a teacher who has read it and speaks of it with enthusiasm. In one indifferent to such a work, we should certainly have little confidence, however he might appear in other respects. Would that every teacher employed in Vermont this winter had the spirit of this book in his bosom, its lessons impressed upon his heart!"—*Vermont Chronicle.*

"I am pleased with and commend this work to the attention of school teachers, and those who intend to embrace that most estimable profession, for light and instruction to guide and govern them in the discharge of their delicate and important duties."—*N. S. Benton, Superintendent of Common Schools, State of New York.*

*Hon. S. Young* says, "It is altogether the best book on this subject I have ever seen."

*President North, of Hamilton College*, says, "I have read it with all that absorbing self-denying interest, which in my younger days was reserved for fiction and poetry. I am delighted with the book."

*Hon. Marcus S. Reynolds* says, "It will do great good by showing the Teacher what should be his qualifications, and what may justly be required and expected of him."

"I wish you would send an agent through the several towns of this State with Page's 'Theory and Practice of Teaching,' or take some other way of bringing this valuable book to the notice of every family and of every teacher. I should be rejoiced to see the principles which it presents as to the motives and methods of good school-keeping carried out in every school-room; and as nearly as possible, in the style in which Mr. Page illustrates them in his own practice, as the devoted and accomplished Principal of your State Normal School."—*Henry Barnard, Superintendent of Common Schools for the State of Rhode Island.*

"The 'Theory and Practice of Teaching,' by D. P. Page, is one of the best books of the kind I have ever met with. In it the theory and practice of the teacher's duties are clearly explained and happily combined. The style is easy and familiar, and the suggestions it contains are plain, practical, and to the point. To teachers especially it will furnish very important aid in discharging the duties of their high and responsible profession."—*Roger S. Howard, Superintendent of Common Schools, Orange Co., Vt.*

*Science of the English Language*

## CLARK'S NEW ENGLISH GRAMMAR.

A Practical Grammar, in which Words, Phrases, and Sentences are classified, according to their offices, and their relation to each other: illustrated by a complete system of Diagrams. By S. W. Clark, A. M. Price 50 cts.

"It is a most capital work, and well calculated, if we mistake not, to supersede, even in our best schools, works of much loftier pretension. The peculiarity of its *method* grew out of the best practice of its author (as he himself assures us in its preface) while engaged in communicating the science to an *adult* class; and his success was fully commensurate with the happy and philosophic design he has unfolded."—*Rahway Register.*

"This new work strikes us very favorably. Its deviations from older books of the kind are generally judicious and often important. We wish teachers would examine it."—*New York Tribune.*

"It is prepared upon a new plan, to meet difficulties which the author has encountered in practical instruction. Grammar and the structure of language are taught throughout by analysis, and in a way which renders their acquisition easy and satisfactory. From the slight examination, which is all we have been able to give it, we are convinced it has points of very decided superiority over any of the elementary works in common use. We commend it to the attention of all who are engaged in instruction."—*New York Courier and Enquirer.*

"From a thorough examination of your method of teaching the English language, I am prepared to give it my unqualified approbation. It is a plan original and beautiful—well adapted to the capacities of learners of every age and stage of advancement."—*A. R. Simmons, Ex-Superintendent of Bristol.*

"I have, under my immediate instruction in English Grammar, a class of more than fifty ladies and gentlemen from the Teachers' Department, who, having studied the grammars in common use, concur with me in expressing a decided preference for 'Clark's New Grammar,' which we have used as a text-book since its publication, and which will be retained as such in this school hereafter."—*Professor Brittan, Principal of Lyons Union School.*

"Clark's Grammar I have never seen equalled for *practicability*, which is of the utmost importance in all school-books."—*S. B. Clark, Principal of Scarborough Academy, Maine.*

"The Grammar is just such a book as I wanted, and I shall make it *the* text-book in my school."—*William Brickley, Teacher at Canastota, N. Y.*

"This original production will, doubtless, become an indispensable auxiliary to restore the English language to its appropriate rank in our systems of education. After a cursory perusal of its contents, we are tempted to assert that it foretells the dawn of a brighter age to our mother tongue."—*Southern Literary Gazette.*

"I have examined your work on Grammar, and do not hesitate to pronounce it superior to any work with which I am acquainted. I shall introduce it into the Mount Morris Union School at the first proper opportunity."—*H. G. Winslow, A. M., Principal of Mount Morris Union School.*

"Professor Clark's new work on Grammar, containing Diagrams illustrative of his system, is, in my opinion, a most excellent treatise on 'the Science of the English Language.' The author has studiously and properly excluded from his book the technicalities, jargon, and ambiguity which so often render attempts to teach grammar unpleasant, if not impracticable. The inductive plan which he has adopted, and of which he is, in teaching grammar, the originator, is admirably adapted to the great purposes of both teaching and learning the important science of our language."—*S. N. Sweet, Author of "Sweet's Elocution."*

www.ingramcontent.com/pod-product-compliance
Lightning Source LLC
LaVergne TN
LVHW020125110826
845151LV00001B/275

* 9 7 8 1 4 2 5 5 4 1 5 9 0 *